*Field Manual
No. 3-90.6

Department of the Army
Washington, DC, 14 September 2010

# Brigade Combat Team

## Contents

*This publication supersedes FM 3-90.6, The Brigade Combat Team, 4 August 2006.

# Figures

# Tables

# Preface

Field manual 3-90.6 provides the commander and staff of the Brigade Combat Team (BCT) and subordinate units with doctrine relevant to Army and joint operations. It applies to the Heavy Brigade Combat Team (HBCT), the Infantry Brigade Combat Team (IBCT), and the Stryker Brigade Combat Team (SBCT). The doctrine described in this manual applies across the full spectrum of military operations – offense, defense, stability or civil support.

This publication:

- Provides BCTs with a framework in which they can operate as part of a division or independently as part of a joint task force.
- Provides doctrine for BCT commanders, staffs, and their subordinate commanders and leaders responsible for conducting major activities performed during operations.
- Serves as an authoritative reference for personnel who –
    - Develop doctrine (fundamental principles and tactics, techniques, and procedures), materiel, and force structure.
    - Develop institution and unit training.
    - Develop unit tactical standard operating procedures for BCT operations.
- Reflects and supports the Army operations doctrine found in FM 3-0, FM 5-0, and FM 6-0.

The proponent of this publication is the U.S. Army Training and Doctrine Command. The preparing agency is the U.S. Army Maneuver Center of Excellence (MCoE), Fort Benning, Georgia. You may submit comments and recommended changes in any of several ways—U.S. mail, e-mail, fax, or telephone—as long as you use or follow the format of DA Form 2028, *Recommended Changes to Publications and Blank Forms*. Contact information is as following:

| | |
|---|---|
| E-mail: | BENN.CATD.DOCTRINE@CONUS.ARMY.MIL |
| Phone: | COM 706-545-7114 or DSN 835-7114 |
| Fax: | COM 706-545-7500 or DSN 835-7500 |
| U.S. mail: | Commander, MCoE |
| | ATTN: ATZB-TDD |
| | Fort Benning, GA 31905-5410 |

This publication applies to the Active Army, the Army National Guard (ARNG)/Army National Guard of the United States (ARNGUS), and the United States Army Reserve (USAR), unless otherwise stated.

Unless otherwise stated in this publication, masculine nouns and pronouns do not refer exclusively to men.

# Summary of Changes

FM 3-90.6 has been revised based on doctrine and terminology changes in FM 3-0. In addition to doctrine changes, technology and organizational revisions have been included.

Another major influence on the development of this manual is an effort by the Combined Arms Doctrine Directorate, U.S. Army Combined Arms Center, to reengineer doctrine by:

- Managing doctrine more effectively.
- Reducing the number of field manuals in order to focus on critical combined arms publications.
- Reducing the size of field manuals to facilitate ease of use, ease of maintenance, and clarity.

As a result, this field manual is considerably shorter than the previous edition. Information found in other doctrinal publications is not repeated but appropriate references are included. Consequently, readers should be familiar with the key field manuals that establish the foundation for the Army's doctrine. These key field manuals are:

- FM 1-02.
- FM 2-0.
- FM 3-0.
- FM 3-07.
- FM 3-13.
- FM 3-28.1.
- FM 3-90.
- FM 5-0.
- FM 6-0.
- FM 6-20.
- FM 6-22.
- FM 7-0.
- FM 7-15.

The significant changes and updates in this new manual include:

- Chapter 1 provides an overview of the role of the BCT, its organization and capabilities, and command and control.
- Chapter 2 describes the types of offensive operations the BCT conducts.
- Chapter 3 includes some changes to terminology and focuses on how a BCT performs defensive tasks.
- Chapter 4 is devoted to stability operations emphasizing its comparable status to offensive and defensive operations. It was restructured to comply with FM 3-0 and FM 3-07 doctrinal changes, counterinsurgency doctrine in FM 3-24, and to include contemporary concepts based on recent Southwest Asia experience.
- Chapter 5 includes minor changes emphasizing a brigade viewpoint of security operations to include an expanded section on area security.
- Chapter 6 includes reconnaissance and surveillance considerations for the BCT. It focuses on intelligence, surveillance, and reconnaissance (ISR) synchronization and integration within the BCT.
- Chapter 7 has been reorganized to conform to doctrinal changes in an upcoming FM that will supersede FM 6-20.

- Chapter 8 covers organizations and activities formerly considered to be combat support, a construct deleted by FM 3-0. It also includes how the BCT uses information engagement and civil-military operations to shape the operational environment. It describes external elements that typically support BCT operations. It also describes doctrinal changes to engineer and site exploitation operations.
- Chapter 9 focuses on the sustainment missions of internal BCT assets, as well as potential augmentation. It also addresses sustainment planning considerations for offensive, defensive and stability operations.

Appendixes, which the previous manual contained, have been deleted.

This page intentionally left blank.

# Chapter 1

# Introduction

Brigade Combat Team (BCT) is a modular organization that provides the division, land component commander (LCC), or joint task force (JTF) commander with close combat capabilities. BCTs are designed for operations encompassing the entire spectrum of conflict. They fight battles and engagements by employing the tactical advantages of a combined arms force structure. BCTs accomplish their missions by integrating the actions of maneuver battalions, field artillery, aviation, engineer, air and missile defense, close air support, and naval gunfire. The BCT's reconnaissance squadron and automated information systems give it information superiority over threat forces. These assets enable the BCT to gather large amounts of information, process it rapidly into intelligence, and disseminate it to decision-makers quickly. This chapter describes the role of the BCT, and how it is organized.

## SECTION I – ROLE OF THE BRIGADE COMBAT TEAM

1-1.   Heavy, Infantry, and Stryker Brigade Combat Teams are the Army's combat power building blocks for maneuver, and the smallest combined arms units that can be committed independently. BCTs conduct offensive, defensive, stability and civil support operations. Their core mission is to close with the enemy by means of fire and maneuver to destroy or capture enemy forces, or to repel enemy attacks by fire, close combat, and counterattack. The BCT can fight without augmentation, but it also can be tailored to meet the precise needs of its missions. BCTs conduct expeditionary deployment and integrate the efforts of the Army with military and civilian, joint and multinational partners.

1-2.   BCTs often operate as part of a division. The division acts as a tactical headquarters that can control up to six BCTs in high- or mid-intensity combat operations, plus a number of supporting functional brigades. The division assigns the BCT its mission, area of operations, and supporting elements, and coordinates its actions with other BCTs of the formation. The BCT might be required to detach subordinate elements to other brigades attached or assigned to the division. Usually the division assigns augmentation elements to the BCT. Fires brigades, battlefield surveillance brigades, maneuver enhancement brigades, sustainment brigades, and aviation brigades can all support BCT operations.

## OPERATIONAL ENVIRONMENT

1-3.   The BCT's unique capabilities (discussed in Section II below) enable strategic and operational military planners to employ the BCT in a variety of conditions, circumstances, and influences. This composite is the operational environment (OE) (FM 1-02). The OE includes all enemy, friendly, and neutral systems across the spectrum of conflict. It also includes an understanding of the physical environment, the state of governance, technology, local resources, and the culture of the local population (FM 3-0).

1-4.   The OE is fluid with continually changing coalitions, alliances, partnerships, and actors. Interagency and joint operations will be required to deal with a wide and intricate range of players occupying the environment. Science and technology, especially information technology, transportation technology, and global economic activity influence the OE. Other trends affect the environment in which the BCT operates. These include demographic changes, movement of populations to urban centers, the global proliferation of electronics and wireless transmissions, climate change, natural disasters, and proliferation of weapons of mass destruction and their effects (FM 3-0).

## OPERATIONAL VARIABLES

1-5. When the BCT is alerted for deployment, redeployment within a theater of operations, or assigned a mission, its higher headquarters provides an analysis of the OE. That analysis includes the operational variables defined in Table 1-1. As a set, these operational variables form the memory aid PMESII-PT. FM 3-0 provides a more detailed description of these operational variables.

**Table 1-1. Operational variables**

| | Variable | Description |
|---|---|---|
| P | Political | The distribution of responsibility and power at all levels of governance. |
| M | Military | The military capabilities of all armed forces in a given operational environment. |
| E | Economic | Individual and group behaviors related to producing, distributing, and consuming resources. |
| S | Social | Societies within the environment. A society is a population whose members are subject to the same political authority, occupy a common territory, have a common culture, and share a sense of identity. |
| I | Information | The aggregate of individuals, organizations and systems that collect, process, disseminate, or act on information. |
| I | Infrastructure | The basic facilities, services, and installations needed to support the local population. |
| P | Physical Environment | Geography and manmade structures. |
| T | Time | Duration of an operation. |

Source: FM 3-0

1-6. The operational variables describe aspects of the environment that are too broad for BCT tactical mission tasks. The BCT commander and staff refine the information about the operational variables and develop mission variables, focusing on those that provide mission-relevant information. Incorporating the operational variables into the mission analysis enhances the BCT commander's and the staff's understanding of the human aspects of the situation (e.g., language, culture, history, education, beliefs) that a mission analysis might otherwise not fully consider. Mission variables are mission, enemy, terrain and weather, troops and support available, time available, and civil considerations (METT-TC) (FM 3-0).

## THREAT IN THE OPERATIONAL ENVIRONMENT

1-7. The BCT must be prepared to deploy anywhere in the world, and to operate against a wide range of threats anywhere within the spectrum of conflict. Some threats come in the form of nation-states. This could be a country or a coalition of countries. Threats can also come from entities that are not states, including insurgents, terrorists, drug traffickers, and other criminal organizations. These non-state actors may use force to further their own interests and threaten the interests of the U.S. or other nation-states. Non-state threats can exist in isolation or in combination with other non-state or nation-state threats.

### Enemy Forces

1-8. The dynamics of warfare between open and complex terrain tactical environments have changed. Recent military operations emphatically demonstrate the value of integrated joint operations. Joint attributes, such as superiority and naval supremacy, enable the BCT to deploy and maneuver freely. To mass combat power, the enemy force needs to maneuver, and when it maneuvers, the enemy force exposes itself to concentrated firepower delivered by joint fires. To avoid exposing his forces, the enemy commander must move his mounted forces with far greater care, seeking cover in towns, villages, and broken ground. Generally, the difficulty of operating mechanized forces without air superiority limits the enemy to dispersed, positional operations, and to limited, local counter attacks along concealed routes.

1-9.   The US military operational concept features close cooperation between highly mobile forces on the ground and air component elements. Because the enemy knows that Army and other air elements have devastating effects against moving mechanized forces in open terrain, he prefers to operate in complex terrain. He makes the best possible use of concealing and covering terrain to avoid exposure to air attack and direct fire engagements. Thus, the BCT must optimize ground tactical formations for operations in complex, rather than open, terrain.

1-10.   In this century, the likelihood of having U.S. forces fighting in urban areas has increased. This is partly because the enemy seeks asymmetric advantages, and partly because the rapid worldwide growth of urbanization makes it difficult to avoid. It is also because of the strategic and operational value of urbanized centers. Cities are vital national resources, and their prompt liberation or seizure can become a political imperative. Moreover, clearing them might become a military, as well as a political, necessity because cities provide sanctuary for vital war-supporting systems, from long-range missiles to command and control (C2). Finally, adversary states or failed states might not choose to, or be able to, oppose U.S. intervention with conventional forces and capabilities but might pursue their strategic aims unconventionally in the challenging terrain of major urban centers. Although the conventional wisdom still might be to defer clearing large urban complexes, strategic necessity often requires land combat forces to enter and control cities.

*Conventional Military Forces*

1-11.   Most of today's conventional military forces have been equipped and organized to meet national needs in regional settings against neighboring states. The U.S. military, with its superior technological, organizational, and strategic capabilities, usually can dominate these regionally focused militaries. However, an alliance of several of these smaller nations could produce a military force capable of challenging the U.S. Army.

*Unconventional Military Forces*

1-12.   Enemies of U.S. policy that do not have capabilities to match the U.S. military must use adaptive methods to achieve their goals. These nations use creative tactics and new technologies to enable them to challenge U.S. forces. Extra-national groups such as insurgents, terrorists, drug traffickers, and other criminal organizations also use unconventional means to oppose U.S. efforts to constrain them.

## Enemy Tactics, Techniques, and Procedures

1-13.   Our enemy's goal is to defeat the BCT's ability to achieve and maintain situational awareness. Towards this goal the enemy employs deception activities and electronic warfare to include:

- Modifying their operations to create false battlefield presentations and reduce signatures through deliberate and expedient means to frustrate intelligence preparation of the battlefield (IPB). The enemy attempts to deceive the BCT by showing it exactly what it expects to see which complicates the process of detecting and assessing threats.
- Positioning decoys and deception minefields in locations where the BCT expects to see them and emplacing real mines where the BCT does not anticipate them.
- Masking the signatures of high-value systems. Differentiating between valid and invalid targets consumes time and affects reconnaissance and surveillance capabilities through deception and dispersion.
- Masking the impact of effects by deception, tampering with indicators, or propaganda to degrade our ability to properly assess the results of operations.
- Exploiting our dependence on wireless networks, electronics, and computer networks by launching electronic warfare attacks.

1-14.   In complex terrain, opponents attempt to deploy undetected by BCT forces. They employ low-signature weapons that are difficult to detect, making protection difficult. This raises the level of uncertainty for moving forces, slowing the pace of BCT maneuver, and thereby making the BCT more vulnerable.

1-15. Enemies seek to complicate BCT targeting by deploying in close proximity to BCT forces, or through shielding forces in cities, among civilian populations, or within landmarks and social or religious structures.

1-16. Enemies conduct operations in a dispersed manner to degrade the BCT's target acquisition and reduce signature. In this operational environment, enemy planning tends to be centralized, and execution is decentralized through coordinated operations. Target effects may be difficult to achieve due to dispersion and signature reduction.

### Enemy in Major Combat Operations

1-17. Enemy forces in major combat operations (MCO) oppose U.S. forces with a variety of means including high-technology niche capabilities built into mechanized, motorized, and light Infantry forces. Possible enemy equipment includes newer generation tanks and Infantry fighting vehicles, significant numbers of antitank guided missile (ATGM) systems, rocket propelled grenades, man-portable air defense (MANPADS) weapons, advanced fixed- and/or rotary-wing aviation assets, missiles, rockets, artillery, mortars, mines, and advanced nonlethal capabilities. The enemy can field large numbers of Infantry and robust military and civilian communications systems. In addition, they may possess weapons of mass destruction. Enemy forces may be capable of long-term resistance using conventional formations such as divisions and corps, as well as sustained unconventional operations and protracted warfare.

### Enemy in Irregular Warfare

1-18. The threat during irregular warfare is likely to be from insurgents, guerrillas, or terrorists. These enemies are highly motivated. They are capable of employing advanced communications and precision weapons (e.g., guided mortar rounds, MANPADS missiles). In addition, they can use ground-based sensors in varying combinations with conventional weapons, mines, and improvised explosive devices (IED). They attempt to shape the information environment to their advantage through such activities as suicide attacks. The threat executes these actions to attract high-profile media coverage or local publicity, and inflate perceptions of insurgent capabilities. Assassinations, kidnappings, and other terrorist acts are common techniques. This type of enemy makes it essential that the BCT uses information engagement with the local population. Whenever possible, the BCT operates in support of host nation government forces, rather than acting as the lead organization. This means task organizing with civil affairs (CA), military information support operations (MISO), and special operations forces (SOF) elements to restore law and order alongside multi-national forces.

## FULL SPECTRUM OPERATIONS

1-19. A BCT operates in a framework of full spectrum operations. FM 3-0 provides a discussion of full spectrum operations, which includes offensive, defensive, and stability or civil support tasks conducted simultaneously. The operational theme under which the joint force operates helps the BCT commander determine the mix of full spectrum operations in which the BCT will participate.

1-20. Offensive operations are combat operations conducted to defeat and destroy enemy forces and seize terrain, resources, and population centers. They impose the commander's will on the enemy. In combat operations, the offense is the decisive element of full spectrum operations. Chapter 2 describes how BCTs conduct the following types of offensive operations:

- Movement to contact.
- Attack.
- Exploitation.
- Pursuit.

1-21. Defensive operations are combat operations conducted to defeat an enemy attack, gain time, economize forces, and develop conditions favorable for offensive or stability operations. Defensive operations can secure and protect areas in which forces conduct stability operations. Defensive operations counter enemy offensive operations. They defeat attacks by destroying as much of the attacking enemy as possible. They also preserve control over land, resources, and populations. Defensive operations retain terrain, guard populations, and protect critical capabilities against enemy attacks. Chapter 3 describes how BCTs conduct, or participate as part of, the following defensive operations:

- Mobile defense.
- Area defense.
- Retrograde.

1-22. Stability operations encompass various military missions, tasks, and activities conducted outside the United States in coordination with other instruments of national power. The goals are to maintain or reestablish a safe and secure environment, provide essential governmental services, emergency infrastructure reconstruction, and humanitarian relief (Joint Publication [JP] 3-0). Forces can conduct stability operations in support of a host nation or interim government, or as part of an occupation when no government exists. Stability operations involve both coercive and constructive military actions. They help to establish a safe and secure environment, and facilitate reconciliation among local or regional adversaries. Stability operations also can help establish political, legal, social, and economic institutions and support the transition to legitimate local governance. Chapter 4 describes how BCTs can perform, or assist in, the following primary stability tasks:

- Civil security.
- Civil control.
- Restoration of essential services.
- Support to governance.
- Support to economic and infrastructure development.

1-23. Civil support is Department of Defense support to U.S. civil authorities for domestic emergencies, and for designated law enforcement and other activities (JP 1-02). Civil support includes operations that address the consequences of natural or manmade disasters, accidents, terrorist attacks, and incidents in the United States and its territories. Army forces conduct civil support operations when the size and scope of events exceed the capabilities or capacities of domestic civilian agencies. The National Guard is suited to conduct these missions; however, the scope and level of destruction may require states to request assistance from Federal authorities. FM 3-28.1 provides more information about these operations.

1-24. BCTs can perform, or support, the following primary civil support tasks:

- Provide support in response to disaster or terrorist attack.
- Support civil law enforcement.
- Provide other support as required.

1-25. The simultaneous conduct of full spectrum operations requires careful assessment, prior planning, and unit preparation as commanders shift their combinations of full spectrum operations. This begins with an assessment of the situation to determine which primary tasks are applicable, and the priority for each. For example, a division assigns a BCT an area of operations and the tasks of eliminating any enemy remnants, securing a dam, and conducting stability operations following a joint offensive phase. The BCT commander determines that the brigade has an immediate primary stability task of establishing civil security—to establish a safe and secure environment in its area of operation and to protect the dam. Simultaneously, the BCT staff begins planning for the next phase in which civil control, and assisting the local authorities with restoring essential services will become priorities while continuing to protect the dam. Reconnaissance and security operations, joint information operations, and protection are continuous. The commander assigns tasks to subordinates, modifies the BCT task organization, replenishes, and requests additional resources if required. Depending on the length of operations, the higher headquarters may establish unit training programs to prepare units for certain tasks.

1-26. When conditions change, commanders adjust the emphasis among the elements of full spectrum operations in the concept of operations. When an operation is phased, these changes are included in the plan. The relative weight given to each element varies with the actual or anticipated conditions. It is reflected in tasks assigned to subordinates, resource allocation, and task organization. Full spectrum operations are not a phasing method. Commanders consider the concurrent conduct of each element—offense, defense, stability and civil support—in every phase of an operation.

## TACTICAL ENABLING OPERATIONS

1-27. The BCT conducts tactical enabling operations to assist the planning, preparation, and execution for the four elements of full spectrum operations (offense, defense, stability, and civil support). Descriptions of the following tactical enabling operations and special environments can be found in the associated references:

- Relief in place (FM 3-90).
- Battle handover (FM 3-90).
- Passage of lines (FM 3-90).
- Linkup (FM 3-90).
- Breaching (FM 3-34.22).
- Gap crossing (FM 3-34.22).
- Clearing (FM 3-34.22).
- Troop movement (FM 3-90).

1-28. The BCT must train these operations to become proficient in their execution. The training and evaluation of these operations should be integrated into field training exercises and other training activities conducted by the BCT that focus on other forms of operations (e.g., offense, defense, stability, civil support).

## SECTION II – ORGANIZATION AND CAPABILITIES

1-29. The BCT is the Army's largest –defined" combined arms organization as well as being the primary close combat force. For combat operations, the ground component of joint task forces is built around the BCT. The BCT includes units and capabilities from every warfighting function; they are task organized to meet specific mission requirements. There are currently three types of BCTs. They are the Heavy Brigade Combat Team (HBCT), the Infantry Brigade Combat Team (IBCT), and the Stryker Brigade Combat Team (SBCT). This section describes the organization, capabilities, and limitations of each BCT type.

1-30. All BCTs include maneuver, fires, reconnaissance, sustainment, military intelligence, military police, signal, and engineer capabilities (Figure 1-1). Higher commanders augment BCTs for a specific mission with capabilities not organic to the BCT structure. Augmentation might include aviation, Armor, cannon or rocket artillery, air defense, military police, civil affairs, military information support operations elements, combat engineers, chemical, biological, radiological, and nuclear (CBRN), and/or additional information systems assets. This organizational flexibility enables BCTs to function across the spectrum of conflict.

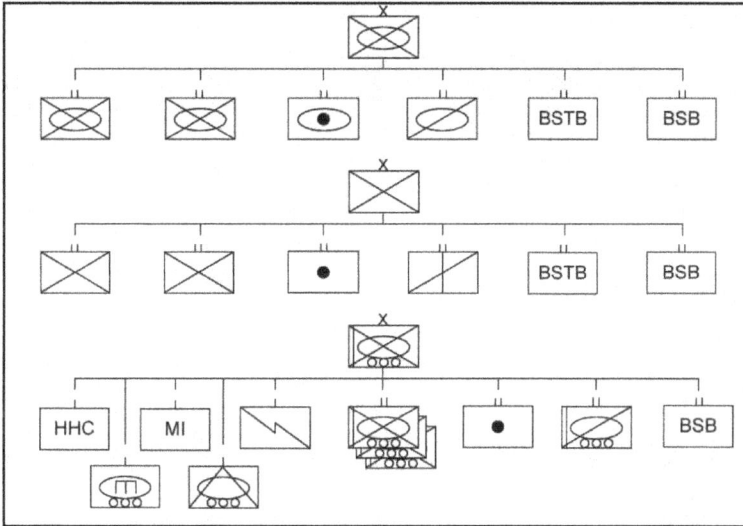

Figure 1-1. Heavy, Infantry, and Stryker BCT organizations

# HEAVY BRIGADE COMBAT TEAM

## ORGANIZATION

1-31. HBCTs are balanced combined arms units that execute operations with shock and speed (Figure 1-2). Their main battle tanks, self-propelled artillery, and fighting vehicle-mounted Infantry provide tremendous striking power. HBCTs require significant strategic airlift and sealift to deploy and sustain. Their fuel consumption may limit operational reach. However, the HBCT's unmatched tactical mobility and firepower offset this. HBCTs include organic military intelligence, military police, artillery, signal, engineer, CBRN, reconnaissance, and sustainment capabilities.

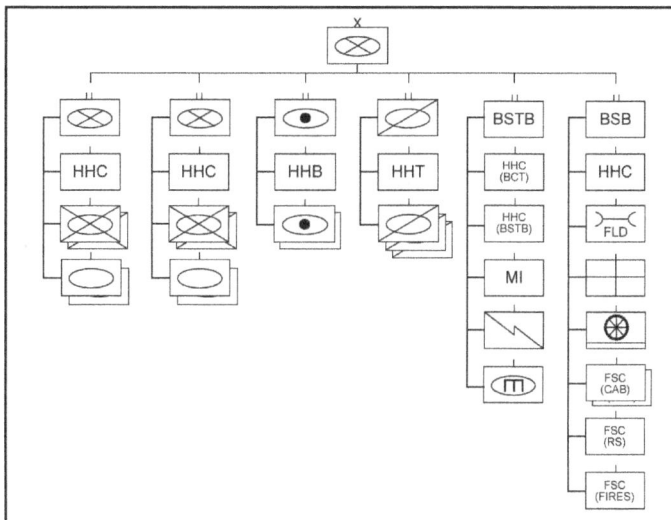

Figure 1-2. Heavy Brigade Combat Team

## Combined Arms Battalion

1-32. The combined arms battalion (CAB) is the HBCTs primary maneuver force. The CAB's mission is to close with, and destroy or defeat enemy forces within the full spectrum of modern combat operations. A CAB maintains tactical flexibility within restricted terrain. It is organized in a "2-by-2" design, consisting of two tank companies and two rifle companies. Companies fight as combined arms teams with support from the CAB's organic 120mm mortars, scout platoon, and sniper squad. Unlike battalions in the IBCT or SBCT, the CAB has a countermine team transporting mine clearing blades and rollers for issue to the tank companies. FM 3-90.5 provides the basic doctrinal principles, tactics, and techniques of employment, organization, and tactical operations appropriate to the CAB.

## Reconnaissance Squadron

1-33. The reconnaissance squadron's fundamental role is to perform reconnaissance. As the –eyes and ears" of the HBCT commander, the reconnaissance squadron provides the combat information that enables the commander to develop situational understanding (SU), make better and quicker plans and decisions, and visualize and direct operations to provide accurate and timely information across the area of operations (AO). It also has the capability to defend itself against most threats. The reconnaissance squadron is composed of four troops: one headquarters and headquarters troop (HHT) and three ground reconnaissance troops equipped with M3 cavalry fighting vehicles (CFV) and armored wheeled vehicles. In large AOs, aerial reconnaissance assets usually are attached or placed under operational control to the squadron to extend its surveillance range. FM 3-20.96 provides the basic doctrinal principles, tactics, and techniques of employment, organization, and tactical operations appropriate to the squadron.

## Fires Battalion

1-34. The HBCT fires battalion provides responsive and accurate fire support including close supporting fires and counterfire. The fires battalion has 16 self-propelled 155-mm howitzers (M109A6 Paladin) in two 8-gun batteries, each with two 4-gun firing platoons. The fires battalion has one AN/TPQ-36, one AN/TPQ-37 counterfire radar and four AN/TPQ-48 lightweight countermortar radars (LCMR). See FM 3-09.70 for additional information on M109A6 Paladin howitzer operations.

## Brigade Support Battalion

1-35. The brigade support battalion (BSB) is the organic sustainment unit of the HBCT. The BSB has four forward support companies that provide support to each of the CABs, the field artillery (FA) battalion and the reconnaissance squadron. These forward support companies provide each battalion commander with dedicated logistics assets (less Class VIII organized specifically to meet the battalion's requirements. The BSB headquarters (HQ) has a distribution management section that receives requests, monitors incoming supplies, and constructs, manages, and distributes configured loads. The BSB also has a supply and distribution company, a field maintenance company, and a medical company assigned to ensure that the HBCT could conduct self-sustained operations for 72 hours of combat.

## Brigade Special Troops Battalion

1-36. The brigade special troops battalion (BSTB) provides command and control to the BCT headquarters and headquarters company (HHC), engineer company, military intelligence company, brigade signal company, military police platoon, and CBRN reconnaissance platoon of the HBCT. It also has a BSTB HHC to provide administrative, logistic, and medical support to its organic and attached units. It is responsible for the security of all BCT command posts (CP) and can, on order, plan, prepare, and execute security missions for areas not assigned to other units in the Brigade area of operations. Its units can defeat small local threats and, with augmentation or control of some of its organic units such as military police, it can organize response forces to defeat threats that are more organized. FM 3-90.61 provides detailed information on the organization and operations of the BSTB.

*Engineer Company*

1-37. The engineer company has three combat engineer platoons and an equipment platoon that enhances assured mobility to the BCT. In addition to limited general engineering tasks, they have the capability to breach and cross obstacles, assist in the assault of fortified positions, emplace obstacles to protect friendly forces, construct or enhance survivability positions, conduct route reconnaissance, and clear improvised explosive devices. It does not have organic gap crossing capability. The engineers also have the mission to maintain the environment for the area of operations.

*Military Intelligence Company*

1-38. The military intelligence company (MICO) mission is to conduct analysis, intelligence synchronization, full motion video, signals intelligence (SIGINT) and human intelligence (HUMINT) collection in support of the BCT and its subordinate commands. The MICO provides analysis and intelligence synchronization support to the BCT S-2. The MICO supports the BCT and its subordinate commands through collection, analysis, and dissemination of intelligence information. It supports the BCT S-2 in intelligence, surveillance, and reconnaissance (ISR) synchronization and in maintaining a timely and accurate picture of the enemy situation.

*Brigade Signal Company*

1-39. The brigade signal company is organic to the HBCT and connects the unit to the global information grid (GIG). The company has a headquarters and two network extension platoons. These platoons consist of a joint network node (JNN) team, a high-capacity line of sight section, a data support team, a wireless network extension team, and an enhanced position location reporting system (EPLRS) network manager and gateway. Usually, one network extension support platoon is located at the BCT main CP, and one at the BSB tactical operations center (TOC). The users supported by the HBCT signal company use the Army Battle Command System (ABCS).

## Heavy Brigade Combat Team Missions, Capabilities, and Limitations

1-40. The following paragraphs provide a discussion of the HBCT mission, its capabilities, and its limitations.

*Heavy BCT Mission*

1-41. The HBCT mission is to fight and win engagements and battles in support of operational and strategic objectives. The HBCT seizes enemy territory, destroys the enemy's armed forces, and eliminates his means of civil population control. The HBCT conducts sustained and large-scale actions in full spectrum operations throughout the depth of the AO. Its combination of firepower, tactical mobility, and organic reconnaissance assets make it invaluable to a higher headquarters commander in combat operations.

*Heavy BCT Capabilities*

1-42. The HBCT's capabilities include:
- Increased firepower, tactical mobility, and protection compared to an IBCT or SBCT, enabling rapid tactical movement, envelopments, and penetrations.
- Conducting full spectrum operations.
- Conducting screen, guard, and cover missions.
- Combined arms integration down to battalion level enabling fighting and winning the close fight.
- Enhanced situational awareness, including a common operational picture (COP), down to individual fighting vehicle level.
- Enhanced linkages to joint forces, fires, and intelligence.
- Robust organic sustainment.

- Accepting and integrating augmented units (e.g., cannon or rocket artillery, air defense, military police, civil affairs, MISO elements, combat engineers).
- Performing company-sized air assaults when augmented with lift assets.
- The BSB has forward support companies for the combined arms and artillery battalions, and the reconnaissance squadron.
- BSTB provides command and control and sustainment support to brigade troops.

### *Heavy BCT Limitations*

1-43. The HBCT's operational limitations include:

- Not rapidly deployable to theater or area of operations (unless deployed to pre-positioned caches).
- Limited capability to conduct forced entry or early entry operations.
- High usage rate of consumable supplies, particularly class (CL) III, V, and IX.
- Possesses no organic gap crossing or general engineering capability, and limited engineer command and control capability.

# INFANTRY BRIGADE COMBAT TEAM

## ORGANIZATION

1-44. The IBCT is the Army's lightest BCT, and is organized around dismounted Infantry, capable of airborne or air assault operations (Figure 1-3). Each of the three types of IBCT (light Infantry, air assault, or airborne) have the same basic organization. IBCTs require less strategic lift and logistical support than other BCTs. When supported with intra-theater airlift, IBCTs have theater-wide operational reach. Organic antitank, military intelligence, artillery, signal, engineer, reconnaissance, and sustainment elements enable the IBCT commander to employ the force in combined arms formations. IBCTs are optimized for operations in close terrain, such as swamps, woods, hilly and mountainous areas, and densely populated areas.

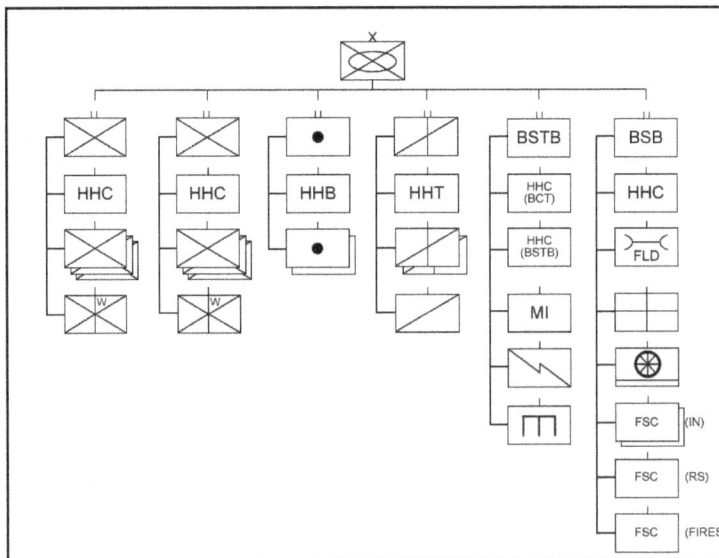

**Figure 1-3. Infantry Brigade Combat Team**

## Infantry Battalion

1-45. Infantry battalions serve as the primary maneuver force for the brigade, and are organized with a HHC, three rifle companies, and a weapons company. Each rifle company has three rifle platoons, a weapons squad, and a 60mm mortar section. The HHC has a scout platoon, a sniper squad, and a platoon of 81mm and 120mm mortars. The weapons company has four wheeled assault platoons, each with three ATGM vehicles. These platoons serve as the IBCT's primary Armor killer providing standoff fires against enemy armor. FM 3-21.20 provides the basic doctrinal principles, tactics, and techniques of employment, organization, and tactical operations appropriate to an Infantry battalion.

## Reconnaissance Squadron

1-46. The reconnaissance squadron of the IBCT is composed of four troops—an HHT, two mounted reconnaissance troops, and one dismounted reconnaissance troop. The two mounted reconnaissance troops are equipped with armored wheeled vehicles. The dismounted reconnaissance troop is easily deployable by either fixed- or rotary-wing aircraft. FM 3-20.96 provides the basic doctrinal principles, tactics, and techniques of employment, organization, and tactical operations specific to the squadron.

1-47. Each of the mounted reconnaissance troops includes three reconnaissance platoons, and a mortar section. The reconnaissance platoons are organized with six armored wheeled vehicles. The mortar section consists of two towed 120mm mortars and a fire direction center (FDC). The dismounted reconnaissance troop includes a sniper squad and two dismounted reconnaissance platoons. The reconnaissance platoons are organized into three sections with one Javelin in each platoon.

## Fires Battalion

1-48. The IBCT fires battalion is organized to provide responsive and accurate fire support including close supporting fires and counterfire. The fires battalion has 16 towed 105-mm howitzers (M119A2) in two 8-gun batteries, each with two 4-gun firing platoons. The fires battalion has 1 AN/TPQ-36 counterfire radar and 4 AN/TPQ-48 LCMRs. FM 3-09.21 provides detailed information on the organization and tactics of the fires battalion.

## Brigade Support Battalion

1-49. The BSB is the organic sustainment unit of the IBCT assigned to ensure that the BCT can conduct self-sustained operations for 72 hours of combat. It is organized as the HBCT BSB described above. The BSB has dedicated troop transportation assets that provide the capability to mount two rifle companies on trucks for a given operation. Each forward support company in a maneuver battalion also has dedicated troop transportation assets that provide the capability to move one rifle company via trucks.

## Brigade Special Troops Battalion

1-50. The IBCT BSTB is organized similarly and provides the same capabilities as the HBCT BSTB described above. FM 3-90.61 provides detailed information on the organization and operations of the BSTB.

## INFANTRY BRIGADE COMBAT TEAM MISSION, CAPABILITIES, AND LIMITATIONS

1-51. The following paragraphs provide a discussion of the IBCT mission, its capabilities, and its limitations.

## Mission

1-52. The IBCT mission is to fight and win engagements and battles in support of operational and strategic objectives. IBCTs can perform complementary missions to HBCTs and SBCTs in offensive operational maneuver. They can be assigned missions such as reducing fortified areas, eliminating enemy force remnants in restricted terrain, securing key facilities and activities, and beginning stability operations in the wake of maneuvering forces. By design, IBCTs are more easily configured for static defensive missions. The IBCT's lack of heavy combat vehicles reduces their logistical footprint. This provides higher commanders greater flexibility in adapting various transportation modes to operationally and tactically move or maneuver the BCT.

1-53. Airborne capable IBCTs conduct airborne assault-specific missions such as forcible entry operations, airfield seizure, expanding an airhead line, and establishing a lodgment. Air assault, another recognized form of forcible entry operation, is a capability common to every IBCT.

### Capabilities

1-54. While IBCTs are optimized for offensive operations against conventional and unconventional forces in rugged terrain, their design also makes them capable in complex terrain defense, urban combat, mobile security missions, and stability operations. IBCTs are better suited for operations in restrictive and severely restrictive terrain than the other types of BCTs. This is true whether the enemy is conventional or unconventional and whether the mission is in support of operational maneuver or operations against insurgents. The IBCT's capabilities include:

- Excellent strategic and operational deployment.
- Capable of conducting forcible entry or early entry operations, including airborne assault, air assault and amphibious operations (Joint Publication [JP] 3-18).
- IBCT is transportable by Army aviation brigades (CH-47 and UH-60 helicopters).
- Enhanced situational awareness, including a common operational picture, down to company commander level (and platoon leaders assigned wheeled vehicles).
- The BSB has forward support companies for the Infantry and artillery battalions, and the reconnaissance squadron.
- BSTB provides command and control and sustainment support to brigade troops.
- Operations require less CL III, V, and IX resupply than HBCT and SBCT.

### Limitations

1-55. Limitations of the IBCT include the following:

- Does not have the organic firepower, tactical mobility, or inherent protection of the HBCT and SBCT.
- No organic gap crossing or general engineering capability, and limited engineer command and control capability.
- The two maneuver battalions of the IBCT move predominately by foot; organic vehicles can transport only two rifle companies at a time.

## STRYKER BRIGADE COMBAT TEAM

### ORGANIZATION

1-56. SBCTs balance combined arms capabilities with significant strategic and intra-theater mobility (Figure 1-4). Designed around the Stryker wheeled armor combat system in several variants, the SBCT has considerable operational reach. It is more deployable than the HBCT and has greater tactical mobility, protection, and firepower than the IBCT. SBCTs fight primarily as a dismounted Infantry formation. The SBCT includes military intelligence, signal, engineer, antitank, artillery, reconnaissance, and sustainment elements. This design lets SBCTs commit combined arms elements down to company level in urban and other complex terrain against a wide range of opponents (FM 3-0).

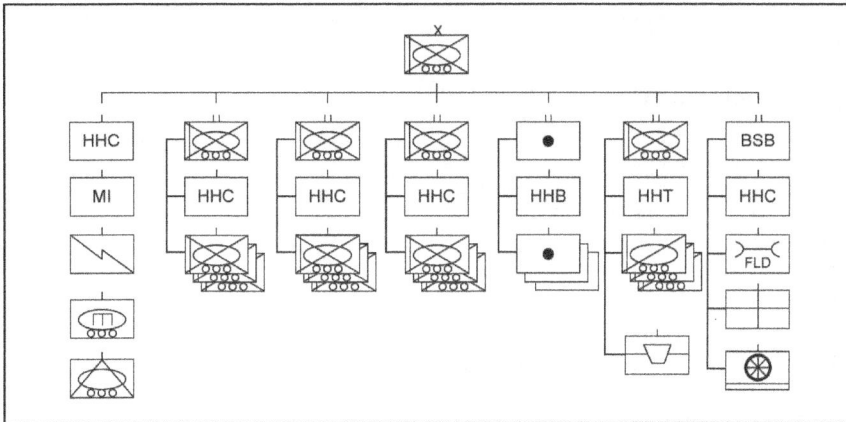

**Figure 1-4. Stryker Brigade Combat Team**

## Infantry Battalion

1-57. SBCT Infantry battalions are organized "3-by-3"; that is, three rifle companies, each with three rifle platoons. Each rifle company has a section of organic 120mm Stryker mortar carrier vehicles with 60mm dismounted mortar capability, a mobile gun system (MGS) platoon with three MGS vehicles, and a sniper team. The HHC also has a mortar platoon equipped with 120mm Stryker mortar carrier vehicles with 81mm mortar dismounted capability, a reconnaissance platoon, and two sniper squads. FM 3-21.21 provides the basic doctrinal principles, tactics, and techniques of employment, organization, and tactical operations appropriate to the SBCT Infantry battalion.

## Reconnaissance Squadron

1-58. The reconnaissance squadron of the SBCT is extremely mobile and can cover a very large area of operations. The reconnaissance squadron is composed of five troops: one HHT, three reconnaissance troops equipped with Stryker reconnaissance vehicles, and a surveillance troop. FM 3-20.96 provides the basic doctrinal principles, tactics, and techniques of employment, organization, and tactical operations appropriate to the squadron.

1-59. Each of the reconnaissance troops includes three reconnaissance platoons and a mortar section. The three reconnaissance platoons contain four reconnaissance vehicles, each with a crew and a scout/HUMINT team for dismounted reconnaissance. The mortar section consists of two 120mm self-propelled mortars and an FDC.

1-60. The surveillance troop provides the squadron commander with a mix of specialized capabilities built around aerial and ground sensors. The tactical unmanned aircraft systems (TUAS) platoon launches, flies, recovers, and maintains the squadron's aerial reconnaissance unmanned aircraft system (UAS) . The multi-sensor ground platoon consists of ground-based radio signals interception and direction-finding teams (e.g., Prophet teams). It also has a dedicated communications terminal that transmits, reports, and receives voice, data, digital, and imagery feeds from intelligence sources at every echelon, from reconnaissance squadron through national level. The ground sensor platoon provides remotely emplaced unmanned monitoring capabilities. The CBRN reconnaissance platoon has three M1135, nuclear, biological, and chemical reconnaissance vehicle (NBCRV) Strykers to determine the presence and extent of CBRN contamination.

## Fires Battalion

1-61. The SBCT fires battalion provides responsive and accurate fire support including close supporting fires and counterfire to the elements of the SBCT. The fires battalion has 18 towed 155-mm howitzers (M777A2) in three 6-gun batteries, each with two 3-gun firing platoons. The fires battalion has 1 AN/TPQ-36 and 1 AN/TPQ-37 counterfire radar and 5 AN/TPQ-48 LCMRs. FM 3-09.21 (which will be revised, renumbered, and renamed) provides detailed information on the organization and tactics of the fires battalion.

### Brigade Support Battalion

1-62. The BSB is the organic sustainment unit of the SBCT. It has four subordinate companies: a distribution company, a forward maintenance company, a medical company, and a HHC. The BSB does not have forward support companies and so must task organize to provide support to each maneuver unit in the SBCT. Chapter 9 of this manual provides more detailed information on the organization and operations of the BSB.

### Antitank Company

1-63. The antitank company is the primary antiarmor force in the SBCT. The company consists of three platoons each with three Stryker ATGM vehicles.

### Engineer Company

1-64. The engineer company provides the SBCT with mobility support. It consists of three mobility platoons and one mobility support platoon. It has limited organic gap crossing capability with four rapidly emplaced bridge systems. The engineers also have the mission to maintain the environment in the area of operations.

### Brigade Signal Company

1-65. The brigade signal company is organic to the SBCT and connects the unit to the GIG. The company has two network extension platoons and various signal support teams under the company headquarters. The users supported by the SBCT signal company use ABCS.

### Military Intelligence Company

1-66. The MICO consists of a headquarters section, an ISR integration platoon, an ISR analysis platoon, and a tactical human intelligence platoon. The ISR analysis and ISR integration platoons are under the operational control (OPCON) of the SBCT S-2; they also provide support to the development of the SBCT COP, targeting effects, situation development, and IPB. They integrate and analyze across the other warfighting functions' reconnaissance and surveillance reporting to develop intelligence products in response to priority intelligence requirements (PIR). The tactical HUMINT platoon provides the SBCT with an organic capability to conduct HUMINT collection and counterintelligence activities.

### STRYKER BRIGADE COMBAT TEAM MISSIONS, CAPABILITIES, AND LIMITATIONS

1-67. The following paragraphs provide a discussion of the SBCT mission, its capabilities, and its limitations.

### Mission

1-68. The SBCT mission is to fight and win engagements and battles in support of operational and strategic objectives. Although the SBCT mainly uses its personnel and equipment to conduct operations other than major combat, when augmented with appropriate heavy Armor capabilities and support, it can execute missions across the full spectrum of conflict. When the SBCT participates in a major combat operation, it does so as a subordinate element of a division or corps. As with any brigade, adjustments to task organization might be required.

### Capabilities

1-69. The SBCT is capable of conducting full spectrum operations, and it provides operational commanders with increased operational and tactical flexibility. This flexibility is enabled by the SBCT's rapid deployment capability (i.e., by air in a matter of days) and by its significantly fewer sustainment requirements (as compared to the HBCT). The SBCT key assets, besides its Soldiers, are its Stryker vehicles and digital information systems (INFOSYS). The Stryker vehicle provides the SBCT both operational and tactical mobility along with added protection and firepower (as compared to the IBCT), while the SBCT's INFOSYS provide enhanced situational awareness down to the vehicle/Soldier level. The SBCT's capabilities include:

- Increased strategic and operational deployment capability compared to a HBCT.
- Capability to conduct air assault forced entry or early entry operations.
- Combined arms integration down to company level.
- Enhanced situational awareness, including a COP, down to individual fighting vehicle level.
- Dismount strength for close combat in urban and complex environments with three Infantry battalions for maneuver (vs. only two in the HBCT and IBCT).
- Limited organic gap crossing capability.
- Lower usage rate of CL III supplies than the HBCT with nearly the same mobility.
- Greater inherent protection than an IBCT.
- Performing company-sized air assaults.

## Limitations

1-70. Limitations of the SBCT include:

- Less firepower or inherent protection than HBCTs.
- Require more intra-theater aircraft to deploy than an IBCT.
- The BSB does not have forward support companies for each maneuver battalion.
- There is no BSTB for C2 of brigade troops.

## SECTION III – BRIGADE COMBAT TEAM COMMAND AND CONTROL

1-71. Command and control (C2) is the exercise of authority and direction by a properly designated commander over assigned and attached forces in the accomplishment of a mission:

- Command is the authority that the BCT commander and subordinate commanders lawfully exercise over subordinates by virtue of rank or assignment.
- Control is the regulation of the BCT and the warfighting functions to accomplish the mission in accordance with the commander's intent.

1-72. The BCT commander performs C2 functions through a C2 system FM 6-0 provides a description of the C2 system – the arrangement of Soldiers, information management, procedures, equipment and facilities essential for the command to conduct operations. An effective C2 system is essential for the BCT commander to conduct (plan, prepare, execute, and assess) operations that accomplish the in operations are battle command and mission command. FM 3-0 provides detailed information on battle command and the BCT commander's role in the operations process. Mission command is the Army's preferred means of battle command.

# BRIGADE COMBAT TEAM MISSION COMMAND

1-73. Mission command is the conduct of military operations through decentralized execution based on mission orders. Mission command rests on the following four elements:

- Commander's intent.
- Individual initiative.
- Mission orders.
- Resource allocation.

1-74. The BCT commander's intent, formalized in the order and understood at company level, provides subordinates with the broad idea behind the operation and allows them to act promptly as the situation requires. Commanders focus their orders on the purpose of tasks and the operation as a whole rather than on the details of how to perform assigned tasks. Orders and plans are as brief and simple as possible.

1-75. Detailed command, in contrast to mission command, centralizes information and decision-making authority. When detailed command is used, orders and plans are comprehensive and explicit. Detailed orders may achieve a high degree of coordination in planning; however, after the operation has begun, they leave little room for adjustment by subordinates and generally fail to remain relevant. Detailed command is not suited for taking advantage of rapidly changing situations and is used by exception. Specific parts of operations often require close coordination in both planning and execution. Examples of such operations include deliberate attack, reliefs in place, passage of lines, air assaults, and major movements.

1-76. There are several considerations the BCT commander makes when he decides whether detailed command is more appropriate. In addition to uncertainty about the situation and the complexity of the operation, the effectiveness of his C2 system—particularly the experience level of his subordinate commanders and staffs—and the amount of time available for planning and preparation are factors to consider. FM 6-0 provides detailed information about mission command.

## OPERATIONS PROCESS

1-77. During operations, the BCT commander balances his time and the staff's time and resources among four major activities in a continuous learning and adaptive cycle called the operations process. The operations process consists of the major C2 activities performed during operations: planning, preparing, executing, and continuously assessing the operation. The commander drives the operations process. FM 3-0 provides detailed information about the operations process. The operations process activities are sequential but not discrete; all overlap and recur as circumstances demand. The commander drives the operations process through battle command.

## INTEGRATING PROCESSES AND CONTINUING ACTIVITIES

1-78. Throughout the operations process, the BCT commanders and staffs synchronize the warfighting functions to accomplish missions. Commanders and staffs use several integrating processes to do this. Where synchronization is the arrangement of action in time, space, and purpose, integration is combining actions into a unified whole. Before commanders can effectively synchronize activities or events, they must integrate activities of the staff. For example, synchronizing fires with movement and maneuver proves difficult if one part of the staff develops the maneuver plan while another part of the staff independently develops the fire plan.

1-79. Integrating processes combine members from across the staff to help synchronize operations. For example, the military decision-making process (MDMP) fosters a shared understanding of the situation as it develops a synchronized plan or order to accomplish a mission. The MDMP not only integrates the actions of the commander, staff, subordinate commanders, and others but also integrates several processes such as IPB, targeting, and airspace command and control (AC2). Commanders and staffs also integrate the warfighting functions through CP cells, working groups, and boards, which are described later in this section.

1-80. The following processes help integrate the staff efforts for particular functions:

- Intelligence preparation of the battlefield. FM 2-01.3 provides information on IPB.
- Intelligence, surveillance and reconnaissance synchronization and integration. FMI 2-01 provides information on ISR synchronization.
- Targeting. FM 6-20-10 provides information on the lethal and nonlethal targeting process.
- Composite risk management. FM 5-19 provides information on risk management.
- Knowledge management. FM 6-0 provides information on knowledge management.
- Airspace command and control. FM 3-52 provides more information on airspace command and control.

1-81. The following activities continue during all BCT operations. They are synchronized with each other and integrated into the overall operation (FM 3-0):

- Reconnaissance and security operations.
- Protection.
- Liaison and coordination.

- Terrain management.
- Information management.
- Airspace command and control.

## Air-Ground Integration

1-82. Operations must be integrated so air and ground forces can simultaneously work in the operational environment to achieve a common objective. Integration maximizes combat power through synergy of both forces. The synchronization of aviation operations into the ground commander's scheme of maneuver may also require integration of other services or coalition partners. It may also require integration of attack reconnaissance, assault, and cargo helicopters.

1-83. Aviation and ground units require effective synchronization and integration to conduct operations successfully and minimize the potential for fratricide and civilian casualties. It continues through planning, preparation, and execution of the operation. To ensure effective integration, commanders and staffs must consider some fundamentals for air-ground integration. The following fundamentals provide the framework for enhancing the effectiveness of both air and ground maneuver assets:

- Understanding capabilities and limitations of each force.
- Use of standard operating procedures (SOP).
- Habitual relationships.
- Regular training events.
- Airspace command and control.
- Maximizing and concentrating effects of available assets.
- Employment methods.
- Synchronization.

1-84. Synchronization is merging the air and ground fights into one with the goal of properly applying aviation capabilities in accordance with the BCT commander's intent. Synchronization ideally begins early in the planning process with brigade aviation element (BAE) involvement. The BAE advises the BCT commander on aviation capabilities and on how to best use aviation to support mission objectives. Of equal importance is ensuring that the BAE/aviation liaison officer (LNO) passes along task and purpose for aviation support, and continually provides updates as needed. Simply stated, ensuring the aviation brigade and subordinate unit staffs fully understand the BCT scheme of maneuver and the commander's intent is critical to successful air-ground integration.

1-85. Employment of attack reconnaissance aviation with ground maneuver forces requires coordinated force-oriented control measures and the close combat attack (CCA) call for fire. This allows aviation forces to support ground maneuver with direct fires while minimizing fratricide risks. The BAE should ensure the BCT is familiar with CCA procedures and marking techniques. FM 3-04.126 contains detailed information about CCA and air-ground integration.

## COMMAND POST OPERATIONS

1-86. The BCT commander organizes his staff into CPs. These CPs provide staff expertise, communications, and information systems that work in concert to aid the commander in planning and controlling operations. All CPs have the responsibility to conduct the five basic functions of information management (IM):

- Collect relevant information.
- Process information from data to knowledge.
- Store relevant information for timely retrieval to support C2.
- Display relevant information tailored for the needs of the user.
- Disseminate relevant information.

1-87. The BCT C2 system consists of the people and equipment in the CPs. The most important element of the C2 system is the people—Soldiers who assist commanders and exercise control on their behalf. Personnel dedicated to the C2 system include seconds in command, command sergeants major, and staffs.

This section describes how the BCT organizes its CPs and their structure. It also describes how the BCT commander cross functionally organizes the staff into CP cells and working groups.

# COMMAND AND CONTROL ORGANIZATIONS

1-88. BCTs are structured to command and control their operations through two command groups and three primary CPs:

- Main CP.
- Tactical command post (TAC CP).
- BSB CP.

1-89. The BCT commander organizes these CPs and command groups by staff sections and staff cells. Staff sections consist of groups of Soldiers by area of expertise (e.g., S-1, S-2, S-3). FM 6-0 provides detailed information about the duties and responsibilities for each Soldier assigned to the BCT staff.

1-90. A cell is a grouping of personnel and equipment according to warfighting function or purpose usually combining a relevant variety of subject matter experts. Organizing the staff among CPs, and into cells within CPs, expands the commander's ability to exercise C2 and makes the system more survivable. Additional staff integration occurs through cross-functional working groups that meet to accomplish specific objectives, typically for producing planning products or assessments. See paragraph 1-130 for more information on BCT working groups.

1-91. Commanders determine the sequence, timing of the deployment or movement, initial locations, and exact organization of CPs based on the situation. Each CP performs specific functions by design as well as tasks the commander assigns. Activities common in all CPs include:

- Maintains running estimates and the COP.
- Controls operations.
- Assesses operations.
- Develops and disseminates orders.
- Coordinates with higher, lower, and adjacent units.
- Conducts knowledge management and information management. See FM 6-01.1 for more information on KM and IM.
- Performs CP administration.
- Maintains local security.

## BRIGADE COMBAT TEAM COMMAND GROUP

1-92. A command group consists of the commander and selected staff members who accompany commanders and enable them to exercise C2 away from a CP. The BCT HQ can form two command groups. They are organized based on the mission. Both are equipped to operate separately from the TAC CP or main CP. Command groups give the commander and the deputy commanding officer (DCO) the mobility and protection to move throughout the AO and to observe and direct BCT operations from forward positions.

1-93. The command group led by the BCT commander consists of whomever he designates. This can include the sergeant major (SGM) and representatives from the S-2, S-3, and fires sections. The command group must have a dedicated security element whenever it departs the main CP. The commander positions his command group near the most critical event, usually with or near the main effort.

1-94. The command group led by the DCO, if used, may include the assistant operations officer, assistant intelligence officer, and a fire support officer. The DCO usually positions his command group with a shaping effort or at a location designated by the BCT commander. The DCO must be able to communicate with the BCT, the battalion commanders, and the CPs. The command group must also have a dedicated security element whenever it departs the main CP.

## BRIGADE COMBAT TEAM MAIN COMMAND POST

1-95. The main CP is the unit's principal CP for BCT C2, internally and externally. It includes representatives of all staff sections and a full suite of INFOSYS to plan, prepare, execute, and assess operations. It is larger in size and staffing, and less mobile than the TAC CP. The BCT executive officer (XO) leads and provides staff supervision of the main CP.

1-96. Functions of the main CP include the following:
- Synchronizing all aspects of decisive, shaping, and sustaining operations.
- Monitoring the current fight.
- Coordinating fires and effects.
- Planning for future operations.
- Monitoring and anticipating commander's decision points and commander's critical information requirements (CCIR).
- Coordinating with higher HQ, adjacent or lateral units and informing them of ongoing missions.
- Supporting the commander's SU through IM.
- Planning, monitoring, and integrating airspace users.
- Develop and implement safety and occupational health, risk management, and accident prevention requirements, policies, and measures.

1-97. The BCT main CP is organized into staff sections, and functional and integrating CP cells to perform specific functions. Commanders adjust their CP organization to fit the situation and their C2 concept for an operation.

1-98. While the main CP can be configured to fit the situation, a typical structure is:
- Current operations cell.
- Plans cell.
- Movement and maneuver cell.
- Fires cell.
- Intelligence cell.
- Protection cell.
- Sustainment cell.

1-99. Considerations for positioning the main CP include:
- Where the enemy can least affect main CP operations.
- Where the main CP can achieve the best communications (digital and voice).
- Where the main CP can control operations best.

> *Note.* In contiguous AOs, the BCT main CP locates behind battalion tactical operations centers and the BCT TAC CP, and out of enemy medium artillery range, if practical. In noncontiguous AOs, the BCT main CP usually locates within a subordinate battalion's AO.

## BRIGADE COMBAT TEAM TACTICAL COMMAND POST

1-100. The TAC CP is a C2 facility containing a tailored portion of the BCT HQ designed to control portions of an operation for a limited time. The BCT commander employs the TAC CP as an extension of the main CP to help control the execution of an operation or specific task, such as a river crossing, a passage of lines, or an air assault operation. The BCT commander may employ the TAC CP to direct the operations of units close to each other when direct command is necessary. This can occur for a relief in place. The commander also can use it to control a special task force, or to control complex tasks such as reception, staging, onward movement, and integration.

1-101. The TAC CP is fully mobile and usually is located near the decisive point of the operation. As a rule, it includes only the Soldiers and equipment essential to the tasks assigned; however, it sometimes requires augmentation for security. The TAC CP relies on the main CP for planning, detailed analysis, and coordination. The BCT XO or the operations officer (S-3) leads the TAC CP. When employed, TAC CP functions include the following:

- Control current operations.
- Provide information to the COP.
- Assess the progress of operations.
- Assess the progress of higher and adjacent units.
- Perform short-range planning.
- Provide input to targeting and future operations planning.
- Provide a facility for the commander to control operations, issue orders, and conduct rehearsals.

1-102. When the TAC CP is not used, the staff assigned to it reinforces the main CP. Unit SOP should address the specifics for this, including procedures to quickly detach the TAC CP from the main CP.

## BRIGADE SUPPORT BATTALION COMMAND POST

1-103. The BSB CP has a special role in controlling and coordinating the administrative and logistical support for the BCT. The improvements in communications and INFOSYS means the BCT no longer has to operate a rear CP collocated with the BSB CP. If necessary, BCT sustainment staff (S-1, S-4, and surgeon), may locate portions of their sections with the BSB CP.

1-104. The BSB CP performs the following functions for the BCT:

- Tracks the current battle so it may anticipate support requirements before units request them.
- In contiguous operations, serves as units' entry point to the BCT's area of operations.
- Monitors main supply routes (MSR) and controls sustainment vehicle traffic.
- Coordinates the evacuation of casualties, equipment, and detainees.
- Coordinates movement of personnel killed in action (KIA).
- Coordinates with the sustainment brigade for resupply requirements.
- Assists in operation of a detainee collection point.
- Provides ad-hoc representation, as required or directed, to the main CP in support of the sustainment cell.

# STAFF

1-105. The staff assists the commander in planning, coordinating, and supervising operations. A staff section is a grouping of staff members by area of expertise under the supervision of a coordinating, personal, or special staff officer. Not all staff sections reside in one of the functional or integrating CP cells discussed later in this chapter. Each staff section maintains its distinct organization. Staff sections operate in different CP cells as required, and coordinate their activities in the various boards, working groups, and meetings established in the unit's battle rhythm (Figure 1-5). FM 6-0 and FM 5-0 provide details about the types and responsibilities of special staff officers.

## EXECUTIVE OFFICER

1-106. The XO is the commander's principal staff officer and the command's principal integrator. He directs staff tasks, manages and oversees staff coordination and special staff officers, and ensures efficient and prompt staff actions. He interacts with the commander's personal staff officers, but he does not necessarily oversee them. The XO usually remains at the main CP.

## COORDINATING STAFF

1-107. Coordinating staff officers are the commander's principal staff assistants. Collectively, through the XO, they are accountable to their commander for all of their responsibilities. Coordinating staff officers advise, plan, and coordinate actions within their areas of expertise. They also exercise planning and supervisory authority over special staff sections as described in FM 5-0.

## PERSONAL STAFF

1-108.   The personal staff sections advise the commander, provide input to orders and plans, and interface and coordinate with entities external to the BCT headquarters. They perform special assignments as directed by the commander. Personal staff officers generally work out of the main CP under the commander's immediate control.

## SPECIAL STAFF

1-109.   Special staff officers help commanders and other staff members perform their responsibilities. The number of special staff officers and their duties vary. Special staff sections are organized according to professional or technical responsibilities. The commander delegates planning and supervisory authority over each special staff function to a coordinating staff officer. Although special staff sections may not be integral to a coordinating staff section, there are usually areas of common interest and habitual association. Special staff officers usually deal routinely with more than one coordinating staff officer. The members of the special staff can change, depending on the situation and the capabilities available to the BCT commander.

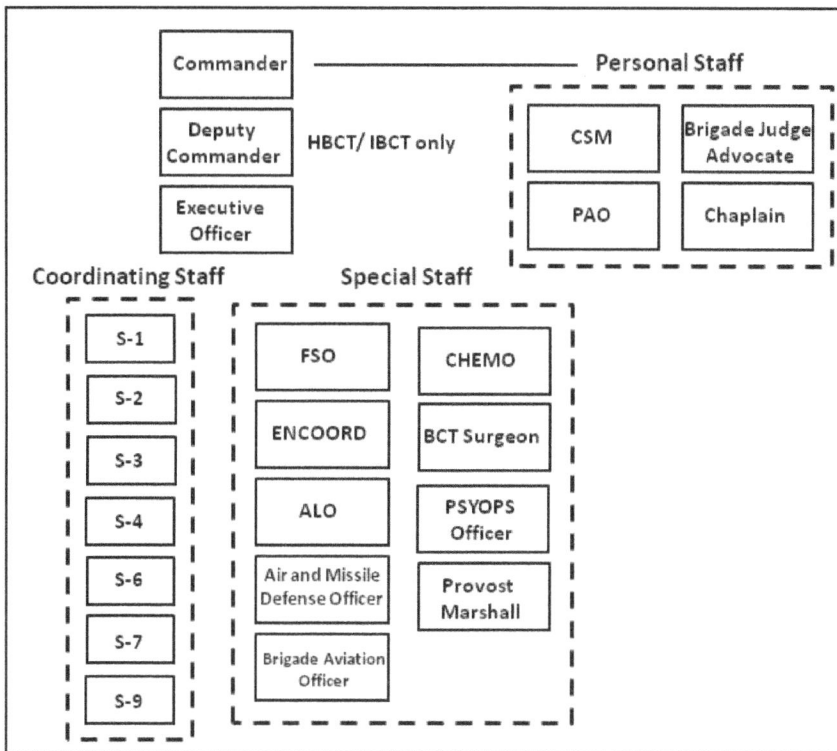

Figure 1-5. BCT staff organization

# CENTERS

1-110.   In addition to CPs, the BCT or its subordinate units may establish centers to assist with coordinating operations. A center is a C2 facility with a supporting staff established for a specific purpose. Centers are similar to CPs in that they are facilities with staff members, equipment, and a leadership component. However, centers have a more narrow focus (for example, a civil-military operations center). Centers usually are formed around a subordinate unit HQ. See Chapter 8 for an example of a civil-military operations center's (CMOC) responsibilities.

# COMMAND POST CELLS AND ELEMENTS

1-111. While each echelon and type of unit organizes CPs differently, two types of CP cells exist: integrating and functional. Integrating cells group personnel and equipment to integrate the warfighting functions according to planning horizon. Functional cells group personnel and equipment according to warfighting function.

## INTEGRATING CELLS

1-112. Cross-functional by design, integrating cells include the plans and current operations cells. The integrating cells coordinate and synchronize forces and warfighting functions within a specified planning horizon. A planning horizon is a point in time that commanders use to focus the organization's planning efforts to shape future events. The three planning horizons are long-range, mid-range, and short-range. Not all echelons and types of units are resourced for all three integrating cells. The CAB, for example, combines their planning and operations responsibilities in one operations cell. The BCT has a small, dedicated planning cell. The BCT is not resourced for a future operations cell. Generally, the BCT focuses on short and mid-range planning, which are associated with the plans cell and current operations cells, respectively. Planning horizons are situation-dependent; they can range from hours and days to weeks and months. As a rule, the higher the echelon, the more distant the planning horizon with which it is concerned.

## PLANS CELL

1-113. The plans cell is responsible for planning operations for the mid- to long-range planning horizons. It prepares for operations beyond the scope of the current order by developing plans, orders, branches, and sequels using the MDMP. The plans cell also oversees military deception planning.

1-114. The plans cell consists of a core group of planners and analysts led by the plans officer. All staff sections assist as required. While the BCT has a small, dedicated plans element, the majority of its staff sections balance their efforts between the current operations and plans cells.

1-115. Upon completion of the initial operation order (OPORD), the plans cell normally develops plans for the next operation or the next phase of the current operation. In addition, the plans cell also develops solutions to complex problems resulting in orders, policies, and other coordinating or directive products such as memorandums of agreement. In some situations, planning teams form to solve specific problems, such as redeployment within the theater of operations. When planning is complete, these planning teams dissolve.

## CURRENT OPERATIONS CELL

1-116. The current operations cell is the focal point for all operational matters. It oversees execution of the current operation. This involves assessing the current situation while regulating forces and warfighting functions in accordance with the commander's intent and concept of operations.

1-117. The current operations cell displays the COP and conducts shift change, assessment, and other briefings as required. It provides information on the status of operations to all staff members and to higher, lower, and adjacent units. The current operations cell has representatives from all staff sections, either permanently or on call. From here, the XO guides the staff and supervises the activities of all cells and staff sections in the main CP.

1-118. The operations officer (S-3) leads the current operations cell from the main CP, or from the TAC CP when separated from the main CP. The movement and maneuver cell forms the core of the current operations cell. Representatives from each staff section and liaison officers from subordinate and adjacent units form the remainder of the cell.

## FUNCTIONAL CELLS

1-119. Functional cells coordinate and synchronize forces and activities by warfighting function. Functional cells within the BCT CP are movement and maneuver, intelligence, fires, sustainment, command, control, communications, and computer operations (part of the C2 warfighting function), and protection. In the BCT, the protection cell function overlaps with fires, movement and maneuver, and C2 warfighting functions.

## Movement and Maneuver Cell

1-120. The movement and maneuver cell coordinates activities and systems that move forces to achieve a position of advantage in relation to the enemy. This includes tasks associated with employing forces in combination with direct fire or fire potential (maneuver), force projection (movement) related to gaining a positional advantage over an enemy, and mobility and countermobility. Elements of the operations, aviation, and engineer staff sections form this cell. The unit's operations officer (S-3) leads this cell. Staff elements in the movement and maneuver cell also form the core of the current operations cell.

## Intelligence Cell

1-121. The intelligence cell coordinates activities and systems that facilitate understanding of the enemy, terrain and weather, and civil considerations. The intelligence cell requests, receives, and analyzes information from all sources to produce and distribute intelligence products. This includes tasks associated with the IPB and ISR synchronization and integration process. Most of the intelligence staff section resides in this cell. The unit's intelligence officer leads this cell.

## Fires Cell

1-122. The fires cell coordinates activities and systems that provide collective and coordinated use of Army indirect fires, joint fires, and C2 warfare through the targeting process. Elements of the fire support (FS), electronic warfare, information engagement and intelligence staff sections make up this cell. The BCT's fire support officer leads this cell.

## Air Defense and Airspace Management/Brigade Aviation Element Cell

1-123. The air defense and airspace management (ADAM)/BAE cell is composed of air defense artillery and Army aviation staff members to coordinate airspace and aviation support issues with other cells. It participates directly in the targeting process and AC2 continuing activity, and may be a part of most working groups and meetings.

## Protection Cell

1-124. The protection cell integrates and synchronizes protection tasks and their associated systems throughout the operations process. The composite risk management process is the overarching process for integrating protection into Army operations. Protection tasks and systems include air and missile defense, personnel recovery, information protection, fratricide avoidance, operational area security, antiterrorism, survivability, force health protection, CBRN operations, safety, operations security, and explosive ordnance disposal. The protection cell coordinates with the network operations cell to facilitate the information protection task. In the BCT, the S-3 supervises the protection cell.

1-125. Protection integration in the BCT may require commanders to designate a staff lead, as the protection officer, who has the experience to integrate risk management and other integrating processes. The executive officer, operations officer (S-3), or a sergeant major could accomplish these duties. Assistant operations officers and other staff officers could be designated as the protection coordinators to facilitate the integration of the twelve protection tasks into operations. In all cases, protection officers and coordinators work with higher and lower echelons to nest protection activities with complementary and reinforcing capabilities.

## Sustainment Cell

1-126. The sustainment cell coordinates activities and systems that provide support and services to ensure freedom of action, extend operational reach, and prolong endurance. It includes those tasks associated with logistics, personnel services, and health service support. Elements of the following staff sections work in the sustainment cell: personnel, logistics, financial management, and surgeon. The BCT S-4 leads this cell.

### Network Operations Cell

1-127. The network operations cell coordinates activities and systems that support communications and information management. Network operations include network management, information dissemination management, and information assurance. The majority of the signal staff section resides in this cell. The BCT command, control, communications, and computer operations officer leads this cell.

# MEETINGS, WORKING GROUPS, AND BOARDS

1-128. In addition to organizing the staff into CP cells and staff sections, the BCT commander may establish meetings, working groups, and boards to integrate the staff and enhance planning and decision-making within the HQ. He may also identify staff members to participate in the higher commander's working groups and boards. The commander establishes and maintains only those working groups and boards required by the situation. The commander, assisted by the XO, establishes, modifies, and dissolves working groups and boards as the situation evolves. The XO manages the timings of these events through the unit's battle rhythm.

1-129. Meetings, working groups, and boards form a major part of a unit's battle rhythm. The XO oversees the battle rhythm and working group scheduling. Each meeting, working group, or board should be sequenced logically so that its outputs are available when needed. The XO balances the time required to plan, prepare for, and hold meetings, working groups, and boards with other staff duties and responsibilities. He also critically examines attendance requirements. Some staff sections and CP cells may lack the personnel to attend all events. The XO and staff members constantly look for ways to combine meetings, working groups, and boards, and eliminate those that are unproductive.

## MEETINGS

1-130. Meetings are gatherings to present and exchange information. They may involve the staff, the commander and staff, or the commander, subordinate commanders, and staff. Cell chiefs and staff section representatives routinely meet to synchronize their activities. Usually meetings that involve the commander end with the commander's guidance. While numerous informal meetings occur daily within a HQ, meetings commonly included in a unit's battle rhythm and the cells responsible for them include:

- Operations synchronization meeting (current operations cell).
- Operations update and assessment briefing (current operations cell).
- Intelligence, surveillance and reconnaissance synchronization meeting (intelligence cell).
- Movement synchronization meeting (sustainment cell).
- Shift change briefing (current operations cell).

## WORKING GROUPS

1-131. A working group is a grouping of designated staff representatives who meet to provide analysis, coordination, and recommendations for a particular purpose or function. Working groups are cross-functional by design and synchronize contributions above the capability of cells and sections. For example, the targeting working group brings together representatives of all staff elements concerned with targeting. It synchronizes the contributions of all staff elements with the work of the fires cell. It also synchronizes fires with current and future operations.

1-132. Typical working groups within the BCT and the lead cell or staff section include:

- Plans working group (plans cell).
- Assessment working group (plans cell).
- Intelligence, surveillance, and reconnaissance working group (current operations cell). See Chapter 6 for more information.
- Targeting working group (fires cell). See Chapter 7 for more information.
- Information engagement working group (fires cell). See Chapter 8 for more information.
- Electronic warfare working group (fires cell). See FM 3-36 for more information.

- Protection working group (protection cell).
- Airspace command and control working group (ADAM/BAE cell). See Chapter 8 for more information.

1-133. Working groups address a number of subjects depending on the situation and echelon. Battalion and brigade HQ usually have fewer working groups than higher echelons. Working groups at battalion and brigade are often less formal. Working groups may convene daily, weekly, or monthly depending on the subject, situation, and echelon.

## BOARDS

1-134. A board is a temporary grouping of selected staff representatives with delegated decision authority for a particular purpose or function. Boards are similar to working groups. However, the commander appoints boards to make or recommend a decision. A board is the appropriate forum when the process or activity needing synchronization requires command approval. Typically, boards address targeting, movements, and assessments.

# INFORMATION SYSTEMS

1-135. INFOSYS, such as the ABCS, substantially enable mission command. They enable the BCT to share the COP with subordinates to guide the exercise of initiative. The COP conveys the BCT commander's perspective and facilitates subordinates' situational understanding. The following information describes the various information systems that support the BCT commander's command and control system.

## ARMY BATTLE COMMAND SYSTEM

1-136. The Army Battle Command System gives the BCT significant advantages in collecting technical information, and distributing information and intelligence rapidly. The ABCS consists of ten core battlefield automated systems plus common services and network management. Each system aids in planning, coordinating, and executing operations by providing access to, and the passing of, information from a horizontally integrated BCT C2 network. The systems that comprise ABCS are (Figure 1-6):

- Tactical Battle Command (TBC). TBC consists of the Maneuver Control System (MCS) and Command Post of the Future (CPOF).
- Global Command and Control System-Army (GCCS-A).
- All Source Analysis System (ASAS). ASAS will be replaced by the Distributed Common Ground System-Army (DCGS-A).
- Battle Command Sustainment Support System (BCS3).
- Air and Missile Defense Workstation (AMDWS).
- Advanced Field Artillery Tactical Data System (AFATDS).
- Force XXI Battle Command-Brigade and Below (FBCB2).
- Tactical Airspace Integration System (TAIS).
- Digital Topographic Support System (DTSS).
- Integrated Meteorological System (IMETS) .
- Integrated System Control [ISYSCON (V) 4].
- Battle Command Common Services (BCCS). The BCCS platform is a collection of server hardware and software application that provides the core ABCS interoperability services and the infrastructure necessary to employ enterprise-class services and an objective service-oriented architecture. BCCS provides CPs at multiple echelons a localized network directory, access control and other services to an expanding array of ABCS and non-ABCS systems (collaboration servers, databases, file servers, websites, email, etc.) and networks that are operated either in a standalone configuration or as part of the GIG.

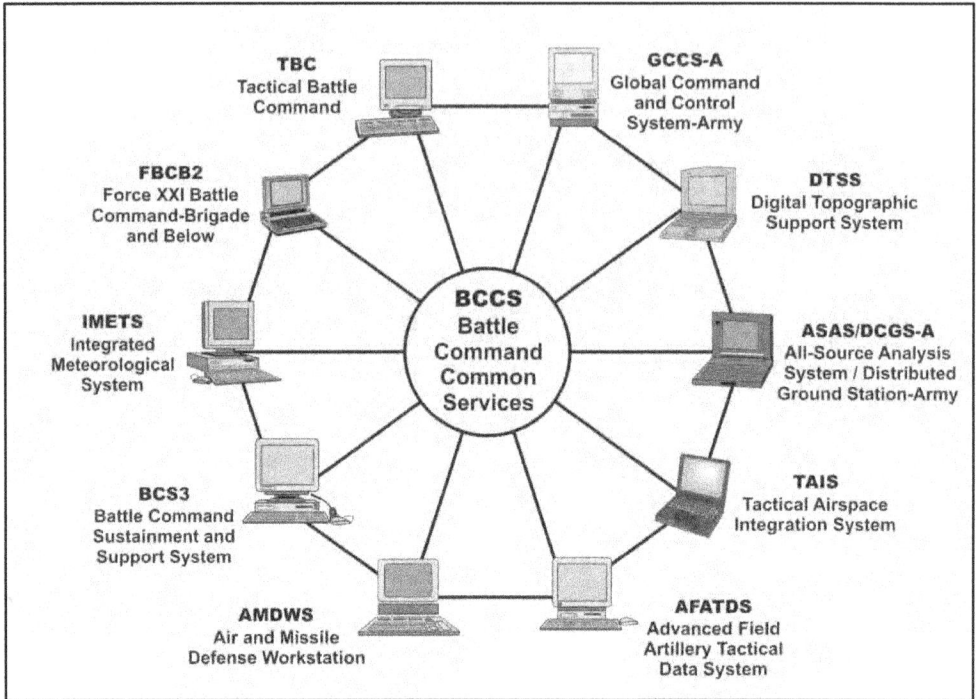

Figure 1-6. Army Battle Command System

## COMBAT NET RADIO AND TACTICAL RADIO SYSTEMS

1-137.  The BCT uses combat net radios (CNR) primarily for voice C2 transmission and secondarily for data transmission where other data capabilities do not exist. The CNR is designed primarily around the single channel ground and airborne radio system (SINCGARS), the single-channel tactical satellite (TACSAT), and the high-frequency (HF) radio.

1-138.  BCTs equipped with blue force tracker (BFT) are capable of communicating between platforms. The BFT system is an L-band satellite communications (SATCOM) tracking and communications system that enables the commander to "see" friendly forces, and provides the commander with the ability to send and receive text messages. BFT is not ABCS-interoperable because it lacks the hardware encrypted secure communications accreditation, which makes BFT noncompatible with EPLRS.

1-139.  BCTs equipped with FBCB2-terrestrial use the EPLRS to provide rapid, jam-resistant, secure data transfer between FBCB2 systems. The EPLRS network provides the primary data and imagery communications transmission system. It is employed in the combat platforms of the commander, executive officer, first sergeant, platoon leaders, and platoon sergeants at the company and platoon level. The EPLRS is an alternate data communications link (host-to-host) between C2 platforms at the brigade and battalion levels. It is the primary data communications link between battalion C2 platforms and company/platoon combat platforms. The EPLRS can be employed in wireless network extension platforms, and configured to provide wireless network extension capability. FM 6-02.53 provides detailed information on the CNR and tactical radio systems.

## JOINT TACTICAL RADIO SYSTEM

1-140.  The joint tactical radio system (JTRS) is the DOD radio of choice for future radio requirements. JTRS will replace the near term digital radio (NTDR). The concept behind the JTRS family of radios is for all services to migrate toward a common type of media among Soldiers, while concurrently out-pacing the growth rate of information exchange requirements, and eventually realizing a fully digitized operational environment. JTRS lays the foundation for achieving network connectivity across the radio frequency (RF) spectrum. The network will provide the means for low to ate digital information exchange, both vertically and horizontally, between warfighting elements. In addition, it will enable connectivity to civil and national authorities.

## INTEGRATED SYSTEM CONTROL

1-141.   The BCT and battalion S-6 sections use ISYSCON to provide communications system network management, control, and planning. The ISYSCON, also known as the tactical internet management system, provides network initialization, local area network management services, and an automated system to support the CNR-based wide area network. ISYSCON features include mission plan management, network planning and engineering, battlefield frequency spectrum management, tactical packet network management, and wide area network  management.

## TRANSPORT SYSTEMS

### Warfighter Information Network-Tactical Increment 1

1-142.   With restructure of the Warfighter Information Network-Tactical (WIN-T) program, both the JNN and WIN-T programs were combined, and JNN is now referred to as WIN-T increment 1 (Inc 1). WIN-T Inc 1 is defined as providing "networking at-the-halt" and is further divided into two sub increments:

- WIN-T Inc 1A. "Extended networking at-the-halt"
- WIN-T Inc 1B. "Enhanced networking at-the-halt"

1-143.   The WIN-T Inc 1 suite of equipment is the network enabler fielded to provide timely, network-enabled support to tactical modular formations. WIN-T Inc 1 provides BCT and battalion connectivity to the GIG. The major components of WIN-T Inc 1 are the hub nodes (fixed, mobile, and tactical); the JNN at corps, division, and brigade; and the CP node at battalion.

1-144.   The JNN enables independent operations and direct termination into the theater network, GIG, or a joint headquarters. The JNN facilitates the management of digital groups, trunks, and circuits. It provides the means through which the communications resource at a node can be monitored, controlled, and managed. The JNN capabilities include Ethernet switching, Internet protocol (IP) routing, network management, and network security services that include network intrusion detection. The JNN works with existing terrestrial transport, high-capacity line-of-sight AN/TRC-190(V3), tropospheric scatter (AN/TRC-170), secure mobile anti-jam reliable tactical–terminal (SMART-T [AN/TSC-154]), and the Trojan SPIRIT II (AN/TSQ-190). FM 6-02.43 provides detailed information on the LandWarNet transport systems at corps and below.

### Brigade Subscriber Node

1-145.   The brigade subscriber node (BSN) is fielded to the BCT as the bridge from MSE to JNN. It provides mobile communications switches and transmission systems through the use of commercial switches and routers. BSN voice subscriber services use a commercial private branch exchange (PBX), which is a private telephone switchboard that allows both integrated service digital network (ISDN) and analog service. Like MSE, individual telephone numbers are assigned to each user. All services are IP automated systems, making the backbone transparent to the user.

This page intentionally left blank.

# Chapter 2

# Offensive Operations

The purpose of the offense is to defeat, destroy, or neutralize the enemy. While the characteristics of offensive operations remain unchanged, the Brigade Combat Team's (BCT) unique capabilities enable it to conduct offensive operations with greater precision and speed than that of past organizations. In past military operations, U.S. ground forces spent precious lives, and consumed extraordinary amounts of munitions and time to develop situations clearly. While costly, this clarity enabled forces to formulate the best solution to the tactical problem. The sensors and information systems (INFOSYS) within the BCT enable the commander to visualize the battlefield better than his predecessors. However, sensors and INFOSYS do not eliminate casualties or render the combined arms assault obsolete. They only enhance the precision and lethality of operations.

## SECTION I – FUNDAMENTALS OF A BRIGADE COMBAT TEAM OFFENSE

2-1.    The brigade combat team conducts offensive operations to defeat and destroy enemy forces and seize terrain, resources, and population centers. Their purpose is to impose United States (U.S.) will on the enemy and achieve decisive victory. Effective offensive operations capitalize on accurate intelligence regarding the enemy, terrain and weather, and civil considerations. Commanders maneuver their forces to advantageous positions before making contact. In offensive operations, the commander seeks to throw enemy forces off balance, overwhelm their capabilities, disrupt their defenses and ensure their defeat or destruction by synchronizing and applying all elements of combat power (FM 3-0).

## CHARACTERISTICS OF THE OFFENSE

2-2.    Surprise, concentration, audacity, and tempo characterize successful offensive operations. BCT commanders sustain the initiative by aggressively committing their forces against enemy weaknesses; BCT attacks are force-oriented or terrain-oriented, and facilitate the defeat of the enemy or the continuation of the attack. BCT commanders extend their attacks in time and space by engaging the enemy in depth, and destroying key elements of the enemy force.

### SURPRISE

2-3.    Commanders achieve surprise by attacking the enemy at an unexpected time or place. The BCT has several capabilities that enable the BCT to achieve surprise. First, the reconnaissance squadron can gain accurate and timely information about the enemy. By visualizing and understanding the situation, the commander can exploit enemy weaknesses and disrupt enemy movement. Second, the movement speed of BCT units either mounted or by air, provides the BCT commander with the option to position combat power rapidly; this limits the enemy's ability to react.

2-4.    The BCT can use its increased information superiority capabilities to give the enemy a false sense of the tactical environment, and possibly lower his defenses. Prior to actual offensive operations, the BCT can use feints and demonstrations to divert the enemy's attention, and tactically deceive him. The key to successful deception is to show the enemy what he expects to see. Often, surprise is achieved by causing the enemy to hesitate physically or in his decision-making. This enables the BCT to retain the initiative by concentrating its forces and adjusting its tempo as the tactical situation requires.

## CONCENTRATION

2-5. Concentration is the massing of overwhelming effects of combat power to achieve a single purpose. The digital communications and INFOSYS found at the company level enable the BCT to concentrate combat power against the enemy quickly. To counter this, the enemy seeks close combat in urban or severely restricted terrain. This terrain inhibits the standoff capability of our weapons systems, makes it difficult to mass effects of combat power with precision, and may cause collateral damage. Using information from higher headquarters (HQ) and joint interagency, intergovernmental, and multinational (JIIM) reconnaissance and surveillance assets, the BCT can create an accurate representation of the terrain and threat forces in its area of operations (AO). This enables the BCT commander to concentrate the BCT reconnaissance efforts on his specific information requirements. This focused effort provides an enhanced COP, which provides the commander and staff a greater situational understanding (SU) than in the past. This SU enables the BCT to concentrate combat power at the decisive point to overwhelm the enemy.

## AUDACITY

2-6. Audacity is a simple plan of action, boldly executed. Since digital capabilities such as Force XXI Battle Command-Brigade and Below (FBCB2) enable the BCT commander to reduce the uncertainties about friendly and enemy forces, the commander can act more boldly.

## TEMPO

2-7. Tempo is the rate of military action. The digital communications and INFOSYS supporting commanders within the BCT enable commanders to process information and disseminate decisions quicker, and to act inside a threat commander's decision cycle.

# SEQUENCE OF OFFENSIVE OPERATIONS

2-8. The commander maneuvers his forces to gain positional advantage so he can seize, retain, and exploit the initiative. He avoids the enemy's defensive strength. He employs tactics that defeat the enemy by attacking through a point of relative weakness, such as a flank, a gap between units, or the rear. For more information on sequencing of offensive operations, refer to FM 3-90. Offensive operations typically follow this sequence:

- **Moving from the assembly area to the line of departure (LD).** The tactical situation and the order in which the commander wants his subordinate units to arrive at their attack positions govern the march formation.
- **Maneuvering from the line of departure to the probable line of deployment (PLD).** Units move rapidly through their attack positions and across the LD, which should be controlled by friendly forces. The commander considers the mission, enemy, terrain and weather, troops and support available, time available, and civil considerations (METT-TC) when choosing the combat formation that best balances firepower, tempo, security, and control.
- **Actions at the PLD, assault position.** The attacking unit splits into one or more assault and support forces as it reaches the PLD, if not already accomplished. All forces supporting the assault should be set in their support by fire position before the assault force crosses the LD. The assault force maneuvers against or around the enemy to take advantage of the support force's efforts to suppress targeted enemy positions.
- **Conducting the breach.** As necessary, the BCT conducts combined arms breaching operations. The preferred method of fighting through a defended obstacle is to employ an in-stride breach. However, the commander must be prepared to conduct deliberate breaching operations. For more information on breaching operations, refer to FM 3-34.22.
- **Assaulting the objective.** The commander employs all means of fire support to destroy and suppress the enemy, and sustain the momentum of the attack. Attacking units move as quickly as possible onto and through the objective. Depending on the size and preparation of enemy forces, it may be necessary to isolate and destroy portions of the enemy in sequence.
- **Consolidating on the objective.** Immediately after a successful assault, the attacking unit seeks to exploit its success. It may be necessary, though, to consolidate its gains. Consolidation can

vary from repositioning force and security elements on the objective, to a reorganization of the attacking force, to the organization and detailed improvement of the position for defense.

- **Transition.** After seizing the objective, the unit transitions to some other type of military operation. This operation could be the exploitation or pursuit, or perhaps a defense. Transitions (through branches and sequels) are addressed and planned prior to the offensive operation being undertaken. Transition operations are discussed at the end of this chapter.

## SECTION II – COMMON OFFENSIVE PLANNING CONSIDERATIONS

2-9. Understanding, visualizing, describing, and directing are aspects of leadership common to all commanders. The BCT commander begins with a designated AO, identified mission, and assigned forces. The commander develops and issues planning guidance based on his visualization in terms of the physical means to accomplish the mission.

2-10. The following discussion uses the warfighting functions (movement and maneuver, intelligence, fires, sustainment, command and control, and protection) as the framework for discussing planning considerations that apply to all types and forms of tactical offensive operations. The commander synchronizes the effects of all warfighting functions as part of the understand, visualize, describe, direct, and assess process.

# MOVEMENT AND MANEUVER

2-11. The commander conducts maneuver to avoid enemy strengths and to create opportunities that increase the effects of his fires. He employs unexpected maneuvers, rapidly changing the tempo of ongoing operations, avoiding observation, and using deceptive techniques and procedures. His security forces prevent the enemy from discovering friendly dispositions, capabilities, and intentions, or interfering with the preparations for the attack. He engages the defending enemy force from positions that place the attacking force in a position of advantage with respect to the defending enemy force, such as engaging the enemy from a flanking position. Finally, he maneuvers to close with and destroy the enemy by close combat and shock effect.

# INTELLIGENCE

2-12. The staff uses METT-TC factors in intelligence preparation of the battlefield (IPB) considerations for offensive operations. Refer to FM 2-01.3 for more information. The staff considers the following when conducting IPB:

- Identify locations of threat/adversary forces, composition, disposition, strengths, and weaknesses of the defending threat/adversary force and their likely intentions, especially where and in what strength the threat/adversary will defend.
- Determine locations of threat/adversary assembly areas, engagement areas, battle positions, indirect-fire weapons system gaps and flanks, electronic warfare (EW) units, and air corridors.
- Determine locations of areas for friendly and threat/adversary air assaults.
- Examine the database to identify how the threat/adversary conducts defensive operations.
- Determine if previous defensive operations are consistent with known threat/adversary doctrine and established threat/adversary models.
- Determine locations of threat/adversary command and control, reconnaissance, and surveillance systems and the frequencies used by the information systems linking these systems.
- Determine a list of intelligence requirements to determine when an enemy force is collapsing so that either an exploitation or pursuit is warranted.
- Determine forecasted weather effect limitations. Commanders need information about weather conditions that affect mobility, concealment, and air operations for both friendly and threat/adversary forces.
- When determining threat/adversary courses of action (COA), war-game as many of the threat/adversary COAs as time permits.

## FIRES

2-13. Units plan and employ fires through a variety of methods and capabilities that attrit, delay, and disrupt enemy forces, and enable friendly maneuver. Using preparatory, counterfire, suppression, and nonlethal fires provides the commander with numerous options for gaining and maintaining fire superiority. The commander uses his long-range artillery systems and air support to engage the enemy throughout the depth of his positions.

2-14. When planning fires in the offense, units should consider the following:
- Position indirect fire assets well forward to exploit weapons ranges and preclude untimely displacement.
- Plan fires for leading elements.
- Plan fires for the neutralization of bypassed enemy combat forces.
- Plan preparation fires, when required, to weaken the enemy's resistance. These fires disrupt, destroy, or damage his defense.
- Plan targets to protect assaulting troops by destroying, neutralizing, or suppressing enemy direct fire weapons.
- Plan fires against enemy reinforcements during the attack and to support friendly consolidations once the objective has been seized.

2-15. The commander establishes his air defense priorities based on his concept of operations, scheme of maneuver, aerial threat, and higher headquarters' priorities. All members of the combined arms team perform air defense operations; however, ground based air defense artillery units execute the majority of the Army's air defense tasks. Air defense assets are allocated to the BCT based on the factors of METT-TC.

## SUSTAINMENT

2-16. The objective of sustainment in offensive operations is to assist the commander in maintaining the momentum. Key to successful offensive operations is the ability to anticipate the requirement to push support forward through a wide dispersion of forces and along lengthy lines of communication (LOC). During offensive operations, certain requirements present special challenges. The most important materiel is typically fuel (Class III bulk) and ammunition (Class V), Class VII, movement control, and medical evacuation. Commanders and staffs must consider establishing aerial resupply and forward logistics bases to sustain operations.

## COMMAND AND CONTROL

2-17. The commander and staff translate the unit's assigned mission into specific objectives for all subordinates, to include the reserve. All planning for offensive operations address the factors of METT-TC, with special emphasis on:
- Enemy positions, strengths, and capabilities.
- Missions and objectives for each subordinate element and task and purpose for each warfighting function.
- Commander's intent.
- AOs for the use of each subordinate element with associated control graphics.
- Consideration of time factors in the operation.
- Scheme of maneuver.
- Special tasks required to accomplish the mission.
- Tactical risk.
- Options for accomplishing the mission.

# PROTECTION

2-18. In offensive operations, the protection cell conducts a risk assessment based on the commander's concept of the operation. This informs the commander of where he assumes risk as he maneuvers to engage the enemy. The staff must recommend to the commander where to focus proactive measures to prevent and deter the threat and mitigate the risk. The level of risk associated with each vulnerability helps to prioritize the application of resources. Key resources to be protected in offensive operations include command posts, lines of communication, and indirect fire assets.

## SECTION III – FORMS OF MANEUVER

2-19. The BCT commander selects the form of maneuver based on his analysis of METT-TC. The commander then synchronizes the contributions of all warfighting functions to the selected form of maneuver. An operation may contain several forms of offensive maneuver, such as frontal attack to clear enemy security forces, followed by a penetration to create a gap in enemy defenses, which in turn is followed by an envelopment to destroy a counterattacking force. The paragraphs below summarize the critical aspects of the forms of maneuver. See FM 3-90 for the characteristics and merits of the forms of maneuver.

2-20. The five forms of maneuver are:
- Envelopment.
- Turning movement.
- Infiltration.
- Penetration.
- Frontal attack.

*Note:* The BCT would normally perform envelopment or turning movement as part of a larger force.

# ENVELOPMENT

2-21. Envelopment seeks to avoid the principal enemy defenses by seizing objectives to the enemy rear to destroy or defeat the enemy in his current positions. At the tactical level, envelopments focus on seizing terrain, destroying specific enemy forces, and interdicting enemy withdrawal routes. In addition, at the tactical level, airborne and air assault operations are vertical envelopments. During an envelopment, the BCT avoids contacting the enemy where the enemy is protected most and has concentrated fires. The reconnaissance squadron and other reconnaissance assets enable the BCT to develop the situation out of contact. The BCT can then maneuver against the enemy on its own terms. Another option is for the BCT to fix the enemy with one force and then attack the enemy with the remaining available force.

# TURNING MOVEMENT

2-22. A turning movement is a form of maneuver in which the attacking force seeks to avoid the enemy's principle defensive positions by seizing objectives to the enemy rear. This causes the enemy to move out of his current positions or divert major forces to meet the threat. This form of offensive maneuver frequently transitions from the attack into an exploitation or pursuit. During a turning movement, the BCT could either serve as a supporting attack, main attack, or reserve force. The division avoids the enemy's principal defensive positions by seizing objectives to the enemy rear, causing the enemy to abandon its prepared defense or to divert major forces to meet the threat. This enables the division to fight the repositioning enemy force on terms and conditions that are favorable to the division (Figure 2-1).

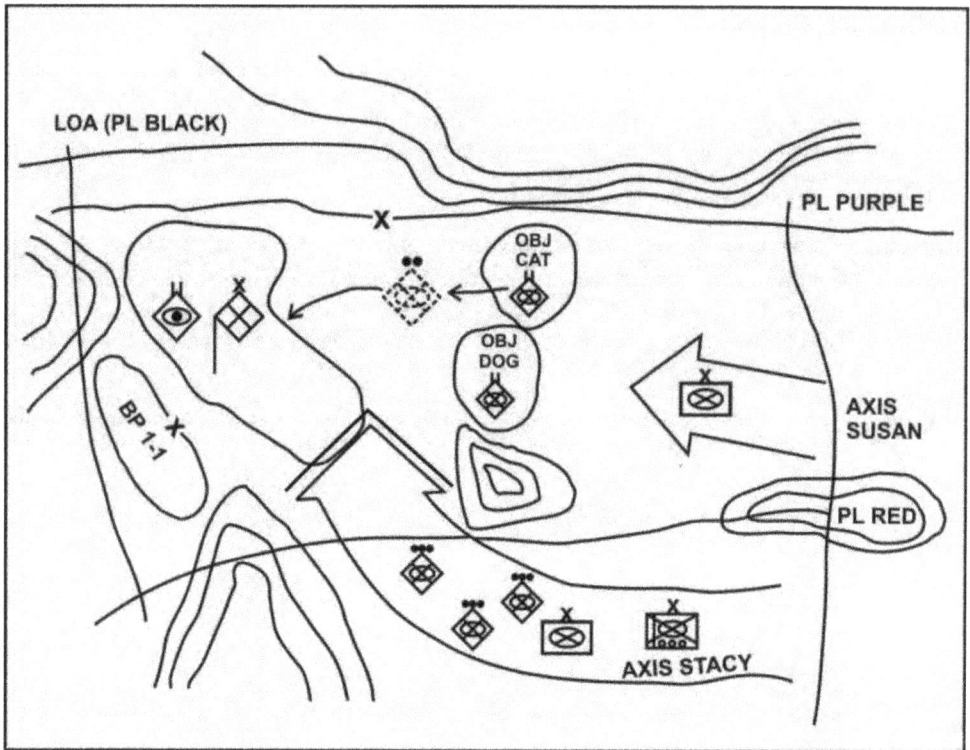

**Figure 2-1. Example of a division conducting a turning movement**

## INFILTRATION

2-23. An infiltration is a form of maneuver in which an attacking force conducts undetected movement through, or into, an area occupied by enemy forces to occupy a position of advantage in the enemy rear while exposing only small elements to enemy defensive fires. The BCT may use infiltration to reconnoiter the enemy force or objective, to attack the enemy from an unexpected location, or to seize terrain to support a future attack. The BCT usually infiltrates reconnaissance assets or Infantry to obtain information or to support the attack by destroying vulnerable key targets or seizing key terrain. Each BCT has the ability to conduct air or ground infiltration. The forces that are used range from reconnaissance to maneuver; and the infiltration can be either mounted or dismounted, depending on specific BCT force structure.

## PENETRATION

2-24. A penetration is a form of maneuver in which an attacking force seeks to breach enemy defenses on a narrow front to disrupt the defensive system. The key to a successful penetration is the ability to mass combat power and effects rapidly at the point of penetration, while maintaining surprise as to the exact location of the penetration. In most cases, the BCT either attacks to create a gap, or attacks through a gap made by another BCT. Figure 2-2 depicts a BCT conducting a frontal assault while a second BCT conducts a deliberate attack.

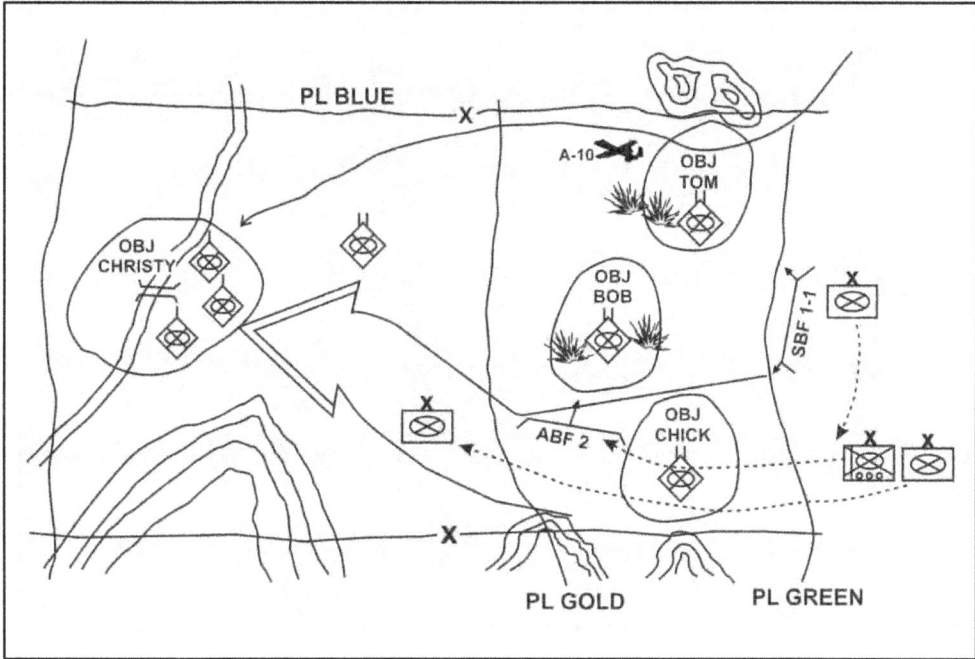

**Figure 2-2. Example of two BCTs conducting a penetration**

# FRONTAL ATTACK

2-25. A frontal attack is a form of maneuver in which an attacking force seeks to destroy a weaker enemy force or fix a larger enemy force in place over a broad front. The BCT can conduct a frontal attack against a stationary enemy or a moving enemy force (Figure 2-3). Depending upon the terrain and enemy, this may not be enough force to execute across a wide front. The IBCT and Heavy Brigade Combat Team (HBCT) must be judicious when deciding to conduct a frontal attack, because they have only two maneuver/combined arms battalions to execute this maneuver. The Stryker Brigade Combat Team (SBCT) generally does not conduct frontal attacks.

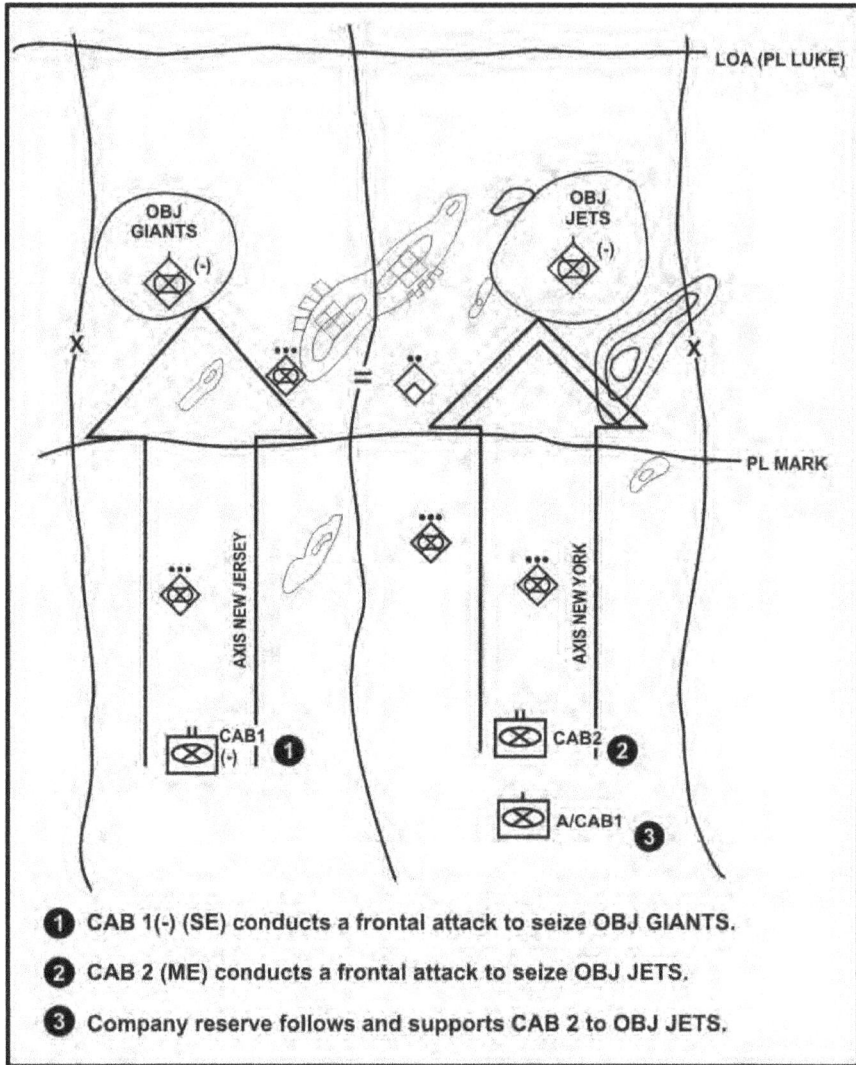

Figure 2-3. Example of a BCT frontal attack against a stationary enemy force

## SECTION IV – PRIMARY OFFENSIVE TASKS

2-26. The BCT conducts, or participates in, movements to contact, attacks, exploitations, and pursuits. The BCT may participate in a division pursuit or exploitation by conducting a movement to contact or attack. The BCT's reconnaissance squadron and reconnaissance assets do not negate the need to conduct the traditional movement to contact. However, the BCT may modify the actual techniques used during movement to contact to fit the capabilities found within the individual BCT.

# MOVEMENT TO CONTACT

2-27. Movement to contact is an offensive operation designed to develop the situation and gain or reestablish contact with the enemy. Units conduct a movement to contact when the tactical situation is unclear or when contact with the enemy has been lost. The purpose is to establish or reestablish direct contact with the enemy. A movement to contact creates favorable conditions for subsequent tactical actions. During a movement to contact, the BCT commander uses his reconnaissance effort to develop the situation, and maintain his freedom of action once contact has been gained. The BCT must also maintain all-around security. The BCT may conduct a movement to contact as part of a higher unit's mission, or the BCT commander may direct a movement to contact any time during an operation. Movements to contact include search and attack, and cordon and search operations.

2-28. The following are fundamentals of a movement to contact:

- Focus all efforts on finding the enemy.
- Make initial contact with organic or JIIM reconnaissance assets or organizations.
- Task organize the BCT to make initial contact with the smallest mobile self-contained force to avoid decisive engagement of the main body.
- Plan to facilitate flexible response throughout the AO.
- Maintain contact once contact is made.

## ORGANIZATION OF A MOVEMENT TO CONTACT

2-29. The minimum components of a BCT movement to contact organization include security forces and a main body.

### Security Forces

2-30. Security forces consist of advance guard, flank and rear security elements.

#### *Advance Guard*

2-31. The advance guard is a self-contained force capable of operating independently of the main body. Only maneuver battalions (i.e., combined arms battalion [CAB] or Infantry battalion) have sufficient combat power to serve as an advance guard. Generally, the advance guard requires field artillery (FA), anti-Armor, and engineer augmentation. In some cases, the division provides a covering force that moves ahead of the BCT. When the division provides a covering force, the BCT commander uses an advance guard to maintain contact with the covering force. The advance guard should remain within range of the main body's indirect fire weapons systems.

#### *Flank and Rear Security*

2-32. When adjacent units are not protecting the BCT's flanks or rear, forces providing a guard or screen secure them. The BCT may use reconnaissance troops for flank security, or it may require the main body forces to provide security.

#### *Reconnaissance*

2-33. The BCT staff develops its intelligence, surveillance, and reconnaissance (ISR) plan to ensure that the appropriate combination of the reconnaissance squadron and other reconnaissance assets are available. The BCT generally focuses its reconnaissance assets where it expects to make contact. Once contact is made, the BCT uses its reconnaissance assets to determine the strength and disposition of enemy forces.

### Main Body

2-34. Once the security forces make contact with the enemy, the BCT commander uses his main body in one of four basic maneuver options.

- Attack.
- Defend and maintain contact.
- Report and bypass.
- Retrograde.

2-35. The main body follows the advance guard and keeps enough distance between itself and the advance guard to maintain flexibility. Depending on his visualization, answers to his commander's critical information requirements (CCIR), and the fidelity of other information, the BCT commander may designate a portion of the main body as the reserve.

## SEARCH AND ATTACK

2-36. Search and attack is one technique for conducting a movement to contact. The BCT conducts this form of movement to contact to destroy enemy forces, deny the enemy certain areas, protect the force, or collect information. Light forces primarily conduct this form of movement to contact. However, heavy forces often support this form of movement. Usually, when the enemy is operating as small, dispersed elements, or when the task is to deny the enemy the ability to move within a given area. The battalion is the echelon that usually conducts a search and attack. The BCT assists its subordinate battalions by ensuring the availability of indirect fires and other support. The BCT may task its subordinate units to conduct the following missions:

- Locate enemy positions or habitually traveled routes.
- Destroy enemy forces within its capability or fix and/or block the enemy until reinforcements arrive.
- Maintain surveillance of a larger enemy force until reinforcements arrive.
- Search urban areas.
- Secure military or civilian property or installations.
- Eliminate enemy influence within the AO.

2-37. Commanders conduct search and attack operations by organizing their units into reconnaissance, fix, and finish forces. Each of these forces has a specific task and purpose. The finish force is the main effort. Some considerations for conducting search and attack operations include intelligence preparation of the battlefield, task organization, isolation of enemy forces, supporting fires, and decentralized command and control (C2).

### Control measures

2-38. The commander establishes control measures that allow for decentralized actions and small-unit initiative to the greatest extent possible. The minimum suggested control measures for a search and attack are an AO, target reference points (TRP), objectives, checkpoints, and coordinating points. The commander can use objectives and checkpoints to guide the movement of subordinate elements. Coordinating points indicate a specific location for coordinating fires and movement between adjacent units. The commander uses other control measures, such as phase lines, as necessary.

### Intelligence Preparation of the Battlefield

2-39. IPB plays a significant role in the planning phase of search and attack missions. The IPB process focuses on the force's reconnaissance effort on likely enemy locations. The intelligence gained from IPB facilitates the BCT's conduct of successful operations through maneuver and fires.

### Task Organization

2-40. In search and attack operations, the commander first task organizes the finish force. He then anticipates the size of the enemy to ensure that the force has enough combat power to accomplish its assigned task. The finish force may move at some distance behind the reconnaissance force, or it may be at a pickup zone (PZ) and air assault to the objective once the enemy is located. The air assault technique is dependent on the availability of landing zones (LZ) near the objective, weather, and availability of aircraft. The S-2 must provide the commander with his estimate of how long it will take the enemy to displace, thus helping to ensure that the finish force reaches the objective before the enemy can displace.

2-41. The size of the reconnaissance force depends on the degree of certainty associated with the enemy template. The vaguer the situation is, the larger the reconnaissance force will be. The reconnaissance force can consist of reconnaissance, Infantry, air, and electronic assets. It usually uses area reconnaissance techniques to reconnoiter named areas of interest (NAI) identified by the S-2. The brigade fix-and-finish plan must consider the possibility of the reconnaissance forces being compromised.

2-42. The BCT can rotate units through the reconnaissance, fix, and finish roles, but the main effort remains with the finish force. Rotating roles may require a change in task organization and additional time for rehearsal.

### Isolate the Enemy

2-43. The fix force isolates the enemy once the reconnaissance force identifies it. It blocks both escape and reinforcement routes. The fix force incorporates indirect fires into the fix plan. It also blocks routes that the S-2 identified as possible escape routes or routes used to reinforce their positions.

### Supporting Fires

2-44. Available fire support (FS) must provide flexible, rapid support throughout the area of operations. This includes the ability to clear fires rapidly. To clear fires rapidly, units must track and report the locations of the unit's subordinate units. The capability must exist to mass fires quickly in support of the main effort. Because of the uncertainty of the enemy situation, the commander avoids command or support relationships that prevent shifting assets when necessary. Supporting fires should be flexible and destructive. They should also enhance the ability of a highly mobile attack force to destroy an enemy force located and fixed by other forces.

### Decentralized Command and Control

2-45. The brigade commander provides the necessary control, but he permits decentralized actions and small-unit initiative to the greatest extent possible within the framework of his intent and the operational concept. This includes establishing the proper graphic control measures to control movement, and synchronization of all brigade assets to enhance combat power.

## CORDON AND SEARCH

2-46. Generally, the BCT commander delegates the cordon and search mission to a battalion (FM 3-06.20). However, the BCT must shape and provide resources for the battalion to accomplish the mission. Subordinate units should divide the built-up area to be searched into zones, and assign a search party to each zone. Search parties consist of:

- Security (or cordon) element–to encircle the area, to prevent entrance and exit, and to secure open areas.
- Search element–to conduct the search.
- Reserve element–to assist either element, as required.

2-47. The BCT can provide the following assets to battalions conducting cordon and search missions:

- Reconnaissance assets from the reconnaissance squadron and the military intelligence company (MICO).
- Mine detection and/or demolition support from attached engineer units.
- Interrogation, translator, and/or human intelligence (HUMINT) support from the MICO.
- Military information support operations (e.g., loudspeaker) and civil affairs (CA) support from attached CA units.
- Electronic warfare support (e.g., Prophet) from the MICO in IBCT/HBCT or reconnaissance squadron in SBCT.
- Liaison officers (LNO) to assist with host nation interaction.
- Host nation security forces.
- Military working dogs and evidence response teams from augmenting military police (MP) units.

### Establishing the Cordon

2-48. An effective cordon is critical to the success of the search effort. Although a battalion usually has sufficient manpower to prevent the escape of individuals to be searched, the BCT may need to allocate forces to protect the battalions conducting the operation. The aviation brigade can provide attack reconnaissance aircraft to support battalions that establish cordons. For further details concerning aviation use in cordon and searches, refer to FM 3-04.126.

2-49. Deployment for the search should be rapid, especially if the enemy is still in the area to be searched. Ideally, the entire area should be surrounded at once. SBCT Infantry battalions have the organic capability to move quickly, but HBCT and IBCT battalions might need additional transport for Soldiers. The BCT should consider the use of the aviation brigade for transport of Soldiers across long distances.

### Conducting the Search

2-50. MISO, CA, and MP units are force multipliers when dealing with the populace. Aerial photographs can provide information needed about the terrain. In larger towns or cities, the local police might have detailed maps showing relative sizes and locations of buildings.

### Capturing Personnel and Equipment

2-51. Commanders must carefully weigh the value of tactically questioning detainees at the point of capture against the thorough questioning by trained interrogators at a safe haven. Although Soldiers on the ground desire to gather and act on timely intelligence, there might be far-reaching damage to an ongoing investigation by military intelligence (MI) or host nation operations. Often MI and host nation representatives can accompany units conducting cordon and search to provide advice to on-site commanders.

2-52. Certain kinds of equipment (e.g., computers and cell phones) or evidence (biometrics or forensic data) may not be exploited effectively at the point of capture. Instead, the BCT should arrange for the collection and quick removal of captured material to MI and other technical experts with the capability to handle exploitation properly.

## HASTY AND DELIBERATE ATTACKS

### HASTY ATTACK

2-53. Hasty attacks maximize agility, surprise, and initiative to seize opportunities to destroy the enemy or seize the initiative. The BCT uses hasty attacks to:

- Exploit a tactical opportunity.
- Maintain the momentum.
- Regain the initiative.
- Prevent the enemy from regaining organization or balance.
- Gain a favorable position that may be lost with time.

2-54. BCTs are capable of conducting more precise hasty attacks because of their inherent battle command enablers such as the Army Battle Command System (ABCS). Rapidly attacking before the enemy can act often leads to success even when the combat power ratio is not as favorable as desired. In choosing to conduct a hasty attack, a commander is trading planning and preparation time for speed of execution. Planning and preparation typically are less detailed for a hasty attack. The BCT can prepare to execute hasty attacks by anticipating their occurrence and developing contingency plans. By assigning on-order or be-prepared missions to subordinate units, the BCT is able to transition into hasty attacks better. Doing so may require a CAB to go from a shaping role to a role as the decisive effort depending on its location during the engagement. In such cases, support and sustainment operations may have to shift to support it. This may require particular attention during the rehearsal phase.

### DELIBERATE ATTACK

2-55. Deliberate attacks are highly synchronized operations characterized by detailed planning, precise preparation, carefully coordinated fires, and violent execution. During shaping operations, BCT commanders allow their reconnaissance effort time to prepare and develop sufficient intelligence to strike the enemy at a vulnerable point with bold maneuver. Shaping operations also disrupt enemy defensive preparations through aggressive patrolling, feints, limited-objective attacks, harassing indirect fires, air strikes, and information engagement tasks. The fires battalion is positioned to provide maximum coverage throughout the initial phases of the operation, and they are prepared to incapacitate the enemy's ability to conduct reconnaissance, conduct strike operations, communicate, and command. Usually, the BCT maneuver/combined arms battalions conduct decisive operations. Fires are planned on all known and potential enemy positions and to isolate the objective. Sustaining operations should be planned to enable rapid resupply and movement. Execution of these operations might require bringing the brigade support battalion (BSB) or some of its crucial elements forward. The reserve is sized to defeat the enemy's counterattack forces. The commander should not constitute his reserve by weakening his decisive operation.

# EXPLOITATION AND PURSUITS

2-56. In an exploitation, the BCT attacks rapidly over a broad front to prevent the enemy from establishing a defense, organizing an effective rear guard, withdrawing, or regaining balance (Figure 2-4). The BCT commander orders pursuit when the enemy can no longer maintain his position and tries to escape. The wheeled mobility of the SBCT, tracked mobility of HBCT, and air capability of the IBCT enable all BCTs to move rapidly into position to conduct the pursuit (Figure 2-5). Successful exploitations and pursuits typically are followed by the capture of enemy prisoners, members of armed groups, or detainees and the BCT must prepare for their rapid disposition.

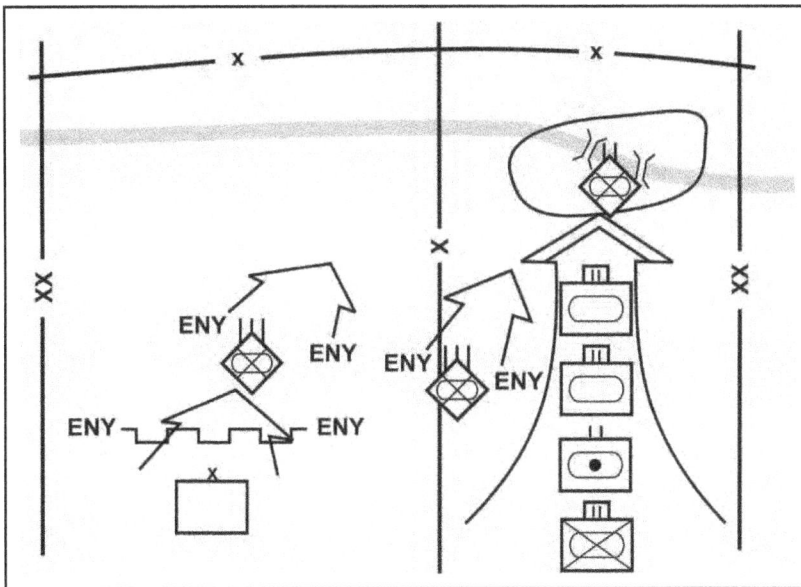

Figure 2-4. HBCT exploitation

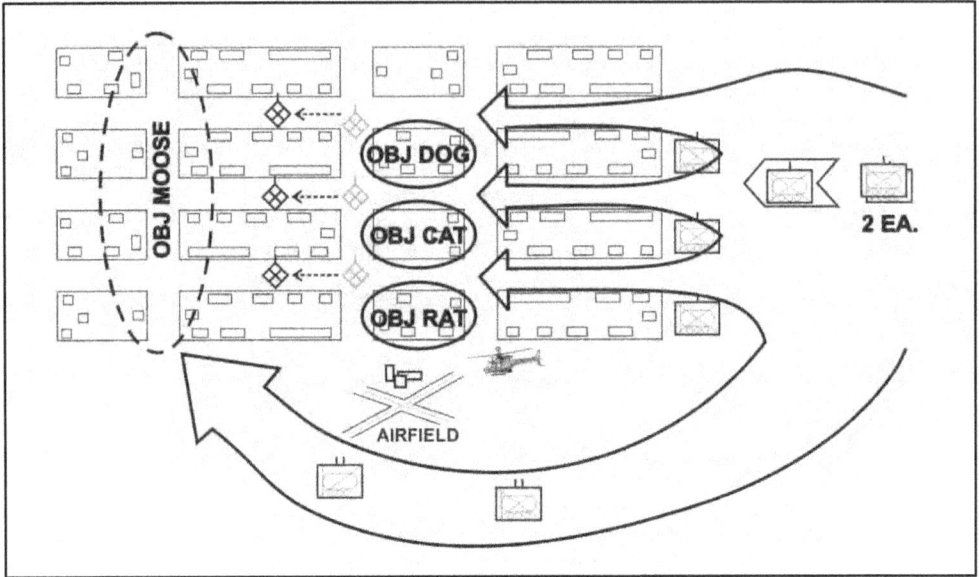

**Figure 2-5. SBCT conducting pursuit**

## SPECIAL PURPOSE ATTACKS

2-57. The BCT can launch attacks with various purposes to achieve different results. The forms of attack that a BCT may conduct are:

- Raids.
- Feints and demonstrations.
- Counterattacks.
- Spoiling attacks.

### RAIDS

2-58. Raids are operations that involve swift penetration of hostile territory to secure information, to confuse the enemy, or to destroy his installations. The BCT makes plans for the withdrawal of personnel at the completion of the mission. The raiding force may operate within or outside of the BCT's supporting range and moves to its objective by infiltration. Once the raid mission is completed, the raiding force quickly withdraws along a different route (Figure 2-6).

2-59. Raids usually are planned at BCT level and executed at battalion level. Battalions usually conduct raids during limited visibility. The approach route should be different from the withdrawal route, which security elements must ensure is open. The raiding force generally carries everything it requires to sustain itself during the operation. If not, resupply is usually by aircraft. Factors that determine the amount of logistics support that must accompany a raiding force include the:

- Type and number of enemy vehicles and weapons.
- Movement distance to the raid objective area.
- Length of time the raid force is to remain in enemy territory.
- Expected enemy resistance.

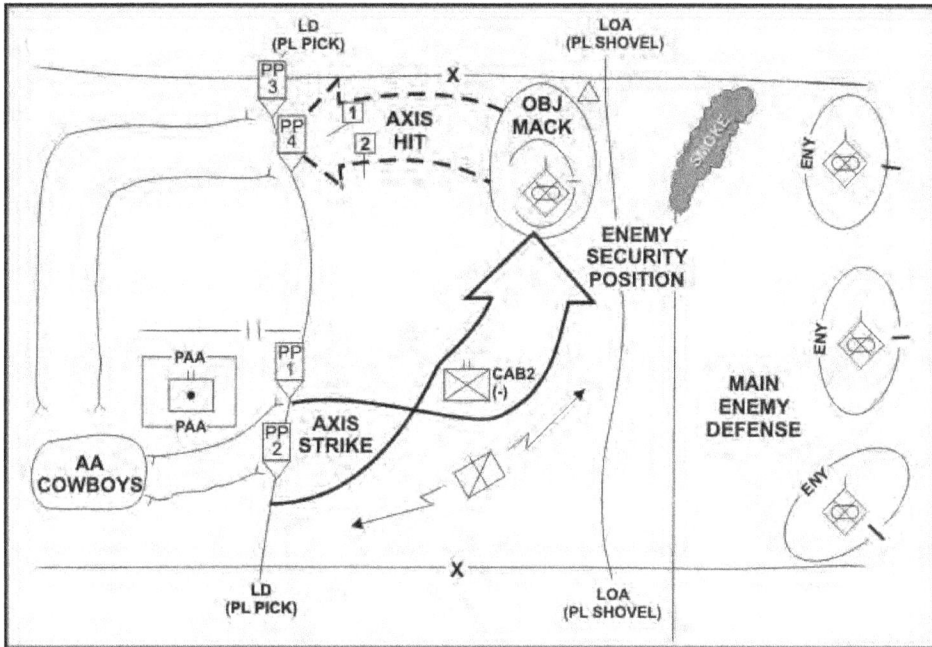

**Figure 2-6. Example of a raid**

## FEINTS AND DEMONSTRATIONS

2-60. A feint is a form of attack used to deceive the enemy as to the location or time of the actual decisive operation. Forces conducting a feint seek direct fire contact with the enemy but avoid decisive engagement. Commanders use feints in conjunction with other military deception activities. They generally attempt to deceive the enemy and induce him to move reserves and shift his FS to locations where they cannot immediately impact the friendly decisive operation or take other actions not conducive to the enemy's best interests during the defense. A demonstration is a form of attack designed to deceive the enemy as to the location or time of the decisive operation by a display of force. Forces conducting a demonstration do not seek contact with the enemy. Both feints and demonstrations are always shaping operations. Demonstrations may be conducted by many elements of the BCT, to include artillery and reconnaissance units. Feints require more combat power and will usually require ground combat units for execution. The BCT commander assigns the operation to the subordinate unit and approves plans to assess the effects generated by it, which may support his operation.

## COUNTERATTACKS

2-61. The commander might plan counterattacks as part of the BCT's defensive plan, or the BCT might be the counterattack force for the higher headquarters. The BCT must resource the counterattack force with enough combat power and mobility to affect the enemy's offensive operations. Figure 2-7 demonstrates a counterattack by the BCT reserve using an attack by fire position to destroy an enemy force while supported by organic artillery.

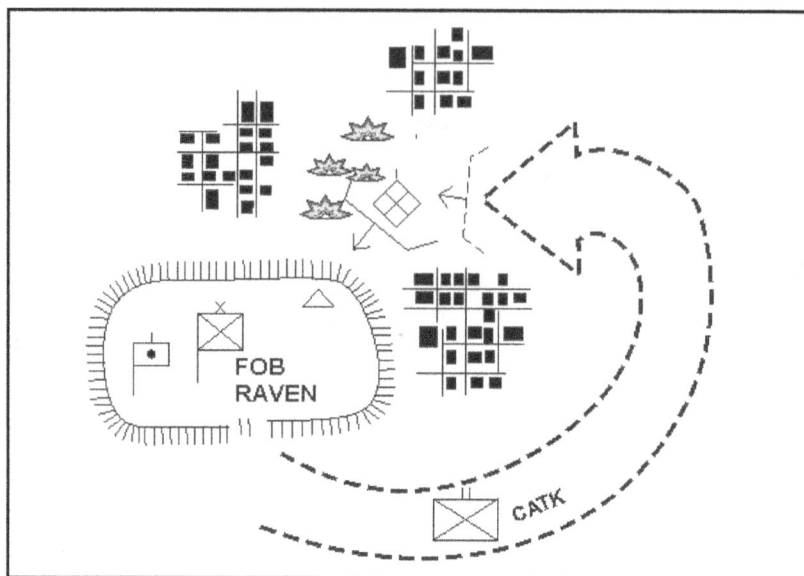

**Figure 2-7. Example of an IBCT counterattack**

SPOILING ATTACKS

2-62. A spoiling attack is a form of attack that preempts or seriously impairs an enemy attack while the enemy is in the process of planning or preparing to attack. The objective of a spoiling attack is to disrupt the enemy's offensive capabilities and timelines while destroying his personnel and equipment—not to secure terrain and other physical objectives (FM 3-90). The BCT commander conducts a spoiling attack, when required, during friendly defensive operations, to strike the enemy while he is in assembly areas or attack positions, preparing for his own offensive operation, or has temporarily stopped. He employs organic reconnaissance and fires elements, as well as JIIM reinforcing elements, to attack enemy assembly positions in front of the friendly commander's main line of resistance or battle positions.

## SECTION V – TRANSITIONS

2-63. Offensive operations are halted when a complete victory is achieved, when a culminating point is reached, or when a change of mission is received from higher headquarters. At this point, the commander transitions his force from one operation to another.

2-64. Following a successful attack, a unit may transition to an exploitation. When possible the lead attacking unit transitions to an exploitation and continues the attack. If this is not feasible, the commander can pass fresh forces (follow and assume) into the lead.

2-65. In order to cut off and destroy enemy forces attempting to retrograde, the unit could transition to a pursuit. The key to success in a pursuit is to apply continuous pressure on a retreating enemy in order to fix the enemy and destroy him. Attacking forces will continue the attack based upon fragmentary orders as the commander and staff readjusts maneuver and sustainment priorities to support the pursuit.

2-66. As offensive operations approach a culmination point, a commander could order a transition to a defense. A commander can use two basic techniques when he transitions to the defense. The first technique is for the leading elements to commit forces and push forward to claim enough ground to establish a security area anchored on defensible terrain. The second technique is to establish a security area generally along the unit's final positions, moving the main body rearward to defensible terrain.

2-67. As offensive operations approach a culmination, or upon order from higher headquarters, a commander could order a transition to a stability operation. These operations establish a safe, secure environment that facilitates reconciliation among local or regional adversaries. Stability operations aim to establish conditions that support the transition to legitimate host-nation governance, a functioning civil society, and a viable market economy. For more information about stability operations, see chapter 4 of this manual or FM 3-07.

## SECTION VI – COMBAT FORMATIONS

2-68. The BCT uses six basic formations: column, line, echelon, box, wedge, and vee. The type of formation the BCT commander selects is based on:

- Planned actions on the objective.
- The likelihood of enemy contact.
- The type of enemy contact expected.
- The terrain the BCT must cross.
- The balance of speed, security, and flexibility required during movement.

2-69. The commander and staff must also determine when, where, and how the BCT transitions into different movement formations based on the terrain and anticipated situation. The commander and all subordinate units also maintain the flexibility to adapt to new formations based on changes in the terrain and enemy situation. FM 3-90 illustrates various organizations using each of these formations.

# COLUMN

2-70. The column formation is useful in restrictive terrain or when attacking on a narrow front. The column formation:

- Is the easiest formation to control.
- Enables rapid movement, especially along roads and trails.
- Provides a high-degree of security and firepower to the flanks.
- Allows follow-on elements to assume the mission or support the lead element (depending on the terrain).
- Provides flexibility for maneuver to the flanks and forward, but is slow to deploy to the front.
- Limits firepower forward.
- Is vulnerable to piecemeal commitment of forces to the front.

# LINE

2-71. The line formation is useful against a weak or shallow enemy defense, or when the situation requires an advance over a broad front. The line formation:

- Provides maximum firepower forward.
- Covers a relatively wide front.
- Facilitates the discovery of gaps, weak areas, and flanks in the enemy's disposition.
- Provides less flexibility of maneuver than other formations.
- Limits firepower to the flanks.
- Requires wide maneuver space for employment and to maintain adequate dispersion.
- Is difficult to control, especially in restricted terrain or during limited visibility.

# ECHELON

2-72. The echelon formation is useful when a BCT flank is threatened, or when maneuvers and enemy contact is expected in the direction of echelon. The echelon formation:

- Enables concentration of firepower forward and to the flank in the direction of echelon.
- Facilitates maneuver against a known enemy in the direction of echelon.
- Allows flexibility in the direction of echelon.

- Transitions easily into a line or vee formation.
- Is easy to control on open terrain but more difficult to control in restricted terrain.
- Requires use of multiple routes or a wide maneuver area.
- Reduces firepower, flexibility of maneuver, and security in the direction opposite of the echelon.

# BOX

2-73. The box formation is useful when general information about the enemy is known, and the BCT requires flexibility and depth in its attack. The diamond formation is a variation of the box formation. The BCT uses box and diamond formations when it has four maneuver forces. Both the box and the diamond formations:

- Provide the best flexibility for maneuver.
- Enable easy transition into all other formations.
- Distribute firepower forward and to the flanks.
- Are easy to control.
- Provide all-around security.
- Facilitate rapid movement.
- Provide protection of accompanying maneuver enhancement and sustainment elements located in the center of the formation.

# WEDGE

2-74. The wedge formation is useful when attacking enemy forces that appear to the front and flank, or when the situation warrants contact with minimal combat power followed by rapid development of the situation. The wedge formation:

- Enables easy transition into other formations.
- Makes contact with minimal combat power forward.
- Provides mutual support between battalions.
- Provides maximum firepower forward and good firepower to the flanks.
- Facilitates control and transition to the assault.
- Is easy to control except in restrictive terrain or during limited visibility.
- Requires sufficient space for lateral and in-depth dispersion.

# VEE

2-75. The vee formation is useful in an advance against a known threat to the front. The vee formation:

- Provides good firepower forward and to the flanks.
- Changes easily to the line, wedge, or column formation.
- Facilitates continued maneuver after contact is made against a relatively weak enemy.
- Is difficult to control, especially in restricted terrain or during limited visibility.
- Is difficult to reorient the formation.

# Chapter 3

# Defensive Operations

During joint operations when a Brigade Combat Team (BCT) is not maneuvering or conducting offensive or stability operations, it conducts defensive operations. Defensive operations are combat operations conducted to defeat an enemy attack, gain time, economize forces, and develop conditions favorable for offensive or stability operations. While the offense is the most decisive type of combat operation, the defense is the stronger type (FM 3-90). Defensive operations are an inherent element and mission throughout full spectrum operations, joint campaigns and homeland security. The defense alone normally cannot achieve a decision. However, it can create conditions for a counteroffensive operation that enables Army forces to regain the initiative. In addition, defensive operations can establish a shield behind which stability operations can progress.

Defensive operations counter enemy offensive operations. They defeat attacks, destroying as much of the attacking enemy as possible. They also preserve control over land, resources, and populations. Defensive operations retain terrain, guard populations, and protect critical capabilities against enemy attacks (FM 3-0). This chapter describes the application of defensive fundamentals to a BCT defense, the primary defensive tasks, defensive planning considerations, and transitioning from the defense to other types of operations.

## SECTION I – FUNDAMENTALS OF A BRIGADE COMBAT TEAM DEFENSE

3-1.  Successful defenses are aggressive. Defending commanders use all available means to disrupt enemy forces. They disrupt attackers and isolate them from mutual support to defeat them in detail. Isolation includes extensive use of command and control warfare. Defenders seek to increase their freedom of maneuver while denying it to attackers. Defending commanders use every opportunity to transition to the offense, even if only temporarily. As attackers' losses increase, they falter and the initiative shifts to the defenders. These situations are favorable for counterattacks. Counterattack opportunities rarely last long. Defenders strike swiftly when the attackers reach their decisive point. Surprise and speed enable counterattacking forces to seize the initiative and overwhelm the attackers.

## CHARACTERISTICS OF THE DEFENSE

3-2.  The BCT uses the preparation time available to create the strongest defense possible. The commander and staff supervise the defensive preparations through inspections and rehearsals. Defensive preparations include:

- Designating a reserve.
- Conducting rehearsals to include rehearsing the reserve and counterattack forces if operational security allows.
- Positioning forces in depth.
- Reinforcing terrain with obstacles.
- Designating, prioritizing, and preparing battle positions and survivability positions.

## RECONNAISSANCE AND SECURITY

3-3. Reconnaissance and security operations seek to confuse the enemy about the location of the BCT's main battle positions (BP), prevent enemy observation of preparations and positions, and keep the enemy from delivering observed fire on the positions. They also force the attacking enemy to deploy prematurely. They can offset the attacker's inherent advantage of initiative regarding the time, place, plan, direction, strength, and composition of his attack by forcing him to attack blindly into prepared defenses. The commander must not permit enemy reconnaissance and surveillance assets to determine the precise locations and strength of defensive positions, obstacles, engagement areas, and reserves. First, the BCT conducts reconnaissance to gain and maintain contact with the enemy. Second, each echelon establishes a security area forward of its main battle area (MBA). All units conduct aggressive reconnaissance and security operations within their area of operations (AO) to seek out and repel or kill enemy reconnaissance and other forces. Units implement operations security measures and information protection to deny the enemy information about friendly dispositions (FM 3-90). See Chapter 5 for more information about security operations.

## DISRUPTION

3-4. The defending force conducts operations throughout the depth of the enemy's formation in time and space. By doing this, the defending force can destroy the enemy's key units and assets, particularly his command and control (C2), artillery and reserves, or disrupt the enemy's timely introduction into battle at the point of engagement. This enables the defending force to regain the initiative. It conducts spoiling attacks to disrupt the enemy's troop concentrations and attack preparations. The defending force uses its reserve, the forces at hand, or a striking force to rapidly counterattack before the enemy can exploit his success. It conducts command and control warfare to assist this process.

3-5. The BCT combines fires, countermobility obstacles including scatterable mines (SCATMINE), defensive positions, and local counterattacks at all levels to disrupt the enemy's attack and break his will to continue offensive operations. Repositioning forces, aggressive local protection measures, and employment of roadblocks, ambushes, checkpoints, and information engagement combine to disrupt the threat of asymmetrical attack.

## MASSING OVERWHELMING COMBAT POWER

3-6. The BCT must mass the effects of its combat power to overwhelm the enemy and regain the initiative. To mass the effects of his forces in the area where he seeks a decision, the commander uses economy of force measures in areas that do not involve his decisive operation. This decisive point can be a geographical objective or an enemy force. In an area defense, defending units use engagement areas to concentrate the effects of overwhelming combat power from mutually supporting positions. Another way he can generate the effects of mass is by committing his reserve (FM 3-90).

3-7. The BCT commander gains situational understanding through the employment of integrated reconnaissance and security operations that answer his information requirements. His situational understanding enables him to shift the effects of fires and maneuver forces so that they repeatedly focus and refocus to achieve decisive, destructive, and disruptive effects upon the enemy's attack. The commander must be audacious in achieving overwhelming combat power at the decisive point while accepting risk, if necessary, in other areas.

## FLEXIBILITY

3-8. The defender gains flexibility through sound preparation, disposition in depth, retention of reserves, and effective C2. Contingency planning permits flexibility. Flexibility also requires that the commander understand and visualize the battlefield to detect the enemy's concept of operations early. Intelligence preparation of the battlefield (IPB) predicts the most likely and dangerous enemy courses of action (COA). Aggressive reconnaissance, counterreconnaissance, and intelligence analysis confirm or deny those actions.

# COMMON PLANNING CONSIDERATIONS

3-9.   Common planning considerations apply to all types of defensive operations (i.e., area, mobile, and retrograde) and focus on several key questions:

- Where is the key and decisive terrain? How can the BCT use key and decisive terrain to defeat/destroy the enemy? Answers to these questions help define the decisive point; that is, where the BCT can best defeat and/or destroy the enemy.
- What conditions must be set to get the enemy to go to the decisive point? This will define shaping operations.
- How will friendly forces and their capabilities combine to bring about synergistic effects? This determines how the engagement area is prepared and how the employment of direct and indirect fires and obstacles will be teamed and synchronized over time or by threat actions.
- How will friendly intentions, plans, and actions be protected and/or portrayed to the enemy (e.g., counterreconnaissance fight and deception operations)?
- Which defensive scheme of maneuver must friendly forces employ? This is answered by analyzing the mission variables with special consideration to terrain, mission (higher commanders' intents), and friendly troops.

## ACHIEVE SITUATIONAL UNDERSTANDING

3-10.   Upon receipt or anticipation of a new mission or a change in mission, the BCT commander makes an initial assessment and develops or refines his situational understanding based upon the specific mission variables or new mission. The outcome of his initial assessment includes:

- Commander's initial guidance.
- Initial timeline.
- Initial warning order.

3-11.   As a part of developing or refining his situational understanding, the BCT commander conducts his own assessment of the mission while his staff conducts its mission analysis. Higher headquarters (HQ) and BCT IPB provide the BCT commander and staff with a clear understanding of how the higher commander envisions the enemy will fight. Key products that help refine the BCT commander's situational understanding derived from mission analysis are:

- Updated running estimates and products.
- Initial IPB.
- Enemy situational templates that address the enemy's most likely and most dangerous COA.
- Modified combined obstacle overlay (MCOO).
- High-value targets (HVT).

## INTELLIGENCE PREPARATION OF THE BATTLEFIELD

3-12.   As with all tactical planning, IPB is a critical part of defensive planning. It helps the commander define where to concentrate combat power, where to accept risk, and where to plan potential decisive actions. To aid in the development of a flexible defensive plan, the IPB must present all feasible enemy COAs. The essential areas of focus are:

- Analyze terrain and weather. How will the enemy use the terrain to his advantage?
- Determine enemy force size; strength; disposition; tactics, techniques, and procedures (TTP)/patterns; morale; and likely COAs with associated defensive positions.
- Determine enemy vulnerabilities and HVTs.
- Determine impact of civilian population on BCT defensive operations.

## HOW AND WHERE TO DEFEAT THE ENEMY

3-13. The BCT commander and staff analyze their unit's role in the higher HQ's fight, and determine how to conduct defensive operations to best achieve their higher commanders' intent. The BCT commander and staff base their determinations of how and where to defeat the enemy on:

- An analysis of the enemy's most likely and most dangerous COAs.
- The location and ability to make use of key and decisive terrain.
- The friendly forces and capabilities available.
- The higher commanders' intents (e.g., conditions at end state).

3-14. In an area defense, the BCT usually achieves success by drawing the enemy into a series of engagement areas. While the enemy is in the engagement area, the BCT destroys it, largely by fires from mutually supporting positions. Most of the defending force is committed to defending positions while the rest is kept in reserve. Commanders use the reserve to maintain the integrity of the defense through reinforcement or counterattack.

3-15. In a retrograde, the BCT achieves success by combining maneuver, fires, obstacles, avoidance of decisive engagement, and operational security until conditions are right to achieve the desired effect. The desired effect includes gaining time, shaping the battlefield for a higher echelon counterattack, withdrawal, or retirement.

## FORCES AND ASSETS AVAILABLE

3-16. The commander and staff analyze the forces and assets available, paying particular attention to the engineer assets and fire support allocated by the higher HQ. The staff must define the engineer and fire support allocation in terms of capability. For example, they define engineer capability in terms of the number of obstacles of a specific effect and the number and type of fighting positions engineers can emplace or create in the time available. Fire support analysis includes the number of targets to be engaged, at what point in the battle they should be engaged, and with what expected result.

3-17. Proper task organization is essential for successful defensive operations. The BCT commander allocates assets where needed to accomplish specific tasks. When developing task organization, the commander must consider all tasks executed during an operation. Changes in task organization may be required to accomplish different tasks during mission execution. Task organizations depend on mission variables.

## MANEUVER

3-18. The BCT can conduct defensive operations with units out of range and/or in mutual support of each other. This requires a judicious effort by the BCT commander and his staff in determining the positioning and priority of support assets/capabilities. During the terrain analysis, the commander and staff must look closely for key and decisive terrain, engagement areas, choke points, intervisibility lines, and reverse slope opportunities in order to take full advantage of the BCT's capabilities to mass firepower in support of defensive maneuver.

3-19. Once the BCT commander has assigned AOs to his maneuver units, he must determine any potential gaps between units. The BCT should plan to cover these gaps with reconnaissance assets. The BCT must plan local counterattacks to isolate and destroy any enemy that manages to penetrate through a gap in the AO. The commander should also plan to reposition units not in contact to mass the effects of combat power against an attacking enemy.

3-20. With the assignment of AOs, the BCT commander also identifies engagement areas where he intends to contain or destroy the enemy force with the massed effect of all available weapons and supporting systems. The commander determines the size and shape of the engagement area by the visibility of the weapons systems in their firing positions and the maximum range of those weapons. The commander designates engagement areas to cover each enemy avenue of approach (AA) into his position (FM 3-90).

3-21. The need for flexibility through the mobility of armored, mechanized, and motorized forces requires the use of graphic control measures to assist in C2 during local counterattacks and repositioning of forces.

Specified routes, phase lines (PL), attack- and support-by-fire positions, BPs, engagement areas, target reference points (TRP), and other fire control measures are required to synchronize maneuver effectively.

## POSITIONING OF THE RESERVE

3-22. Positioning the reserve is critical to effective employment. The reserve requires adequate depth to have a degree of protection and to prevent inadvertent commitment too early in the fight. However, the reserve must be close enough that it can rapidly enter the fight when committed. The reserve can occupy battle, blocking, or hide positions.

### Fire Support

3-23. The following are considerations for the fire support plan:

- Allocate initial priority of fires to the security force.
- Plan targets along enemy reconnaissance mounted and dismounted AAs.
- Engage approaching enemy formations at vulnerable points along their route of march with indirect fires and close air support (CAS) if available.
- Plan the transition of fires to the MBA fight.
- Develop clear triggers to adjust fire support coordinating measures (FSCM) and priority of fires.
- Ensure integration of fires in support of obstacle effects.
- Ensure integration of fires with BCT counterattack plans and repositioning contingency plans.
- Integrate the emplacement of SCATMINEs into the countermobility and counterattack plans.

### Engineer Support

3-24. The priority of effort to transition from mobility to countermobility and survivability requires detailed planning at the BCT level to ensure subordinate engineers have adequate time for troop leading procedures. Engineer augmentation provides survivability support to the BCT. The engineer coordinator (ENCOORD) and supporting combat engineer company commander are key in the development and execution of engineer tasks. The following planning considerations apply to engineer support:

- Position situational obstacles early and link them to natural and other manmade obstacles.
- Plan multiple obstacle locations to support depth and flexibility in the defense. Ensure adequate security for obstacle emplacement systems. Integrate triggers for the execution of situational and reserve obstacles in the decision support template (DST).
- Focus the countermobility effort to cause the enemy to maneuver into positions of vulnerability where the BCT intends to kill them.
- Ensure adequate mobility support for withdrawing security forces, the reserve, the counterattack force, and the repositioning of MBA forces.
- Ensure the integration of survivability priorities for critical systems and units through the development and implementation of an execution matrix and timeline.

### Protection Operations

3-25. Air and missile defense (AMD) support to the BCT may be limited. Units should expect to use their organic weapons systems for self-defense against enemy air threats. Units should plan chemical, biological, radiological, and nuclear (CBRN) reconnaissance at likely locations in anticipation of possible enemy employment of CBRN agents and hazards. Use obscurants to support disengagement or movement of forces. Assign sectors of fire to prevent fratricide.

### Aviation Support

3-26. In defensive operations, the speed and mobility of aviation can help maximize concentration and flexibility. During preparation for defensive operations, aviation units sometimes support the BCT commander with aerial reconnaissance and fires.

3-27. Attack reconnaissance helicopters routinely support security area operations and mass fires during the MBA fight. Synchronization of aviation assets into the defensive plan is important to ensure aviation assets are capable of massing fires and to prevent fratricide. If the BCT is augmented with aviation assets, it must involve the direct fire planning processes of the supporting aviation unit, through its aviation liaison officer (LNO) and the ADAM/BAE cell.

**Sustainment Support**

3-28. The BCT logistics staff officer (S-4) must ensure that the sustainment plan is fully coordinated with the rest of the staff. He coordinates with the operations staff officer (S-3) to ensure that supply routes do not interfere with maneuver or obstacle plans but still support the full depth of the defense. The S-4 coordinates with the CBRN officer to ensure there are appropriate routes for contaminated equipment. In addition, the S-4 coordinates with the forward support company commander for the possible use of pre-stocked classes of supply (class [CL] IV and V).

3-29. Enemy actions and the maneuver of combat forces complicate forward area medical operations. Health service support (HSS) considerations for defensive operations include:

- Medical personnel have much less time to reach the patient, complete vital emergency medical treatment, and remove the patient from the battle site.
- The enemy's initial attack and the BCT's counterattack produce the heaviest patient workload. These are also the most likely times for enemy use of artillery and CBRN weapons.
- The enemy attack can disrupt ground and air routes and delay evacuation of patients to and from treatment elements.
- The depth and dispersion of the defense create significant time-distance problems for evacuation assets.

3-30. The enemy exercises the initiative early in the operation, which could preclude accurate prediction of initial areas of casualty density. This fact makes effective integration of air assets into the medical evacuation (MEDEVAC) plan essential.

### CIVILIAN CONCERNS

3-31. Generally, the land component commander or the joint forces commander establishes rules of engagement (ROE) to avoid civilian casualties and damage to medical facilities, historical and cultural sites, and critical civilian infrastructure. These civil considerations are analyzed in terms of relevant areas, structures, capabilities, organizations, people, and events. ROE might restrict use of cluster munitions, mines, nonlethal gas, obscurants, and even mortar fires. The rules might prohibit firing into towns or in the vicinity of refugees. The BCT should plan for protection and removal of civilians who enter battle areas.

# COMMON CONTROL MEASURES

3-32. For an area or retrograde type of defense, the BCT organizes its defensive forces, areas, and actions under a framework that includes a security area, an MBA, and a reserve. These categories describe for BCT subordinate elements both a physical area and a tactical intent. BCT elements may transition from one part of the framework to another during the operation, such as completing a security mission in the security area and then moving to defend a BP in the MBA or occupying an assembly area and assuming a reserve role. Mission, enemy, terrain and weather, troops and support available, time available, and civil considerations (METT-TC) analysis influences the physical areas and tactical responsibilities for BCT elements within this framework.

*Note.* This framework construct applies to an area defense and retrograde operations. BCTs do not normally conduct a mobile defense. This is because of their inability to fight multiple engagements throughout the width, depth, and height of the AO while simultaneously resourcing striking, fixing, and reserve forces. Typically, the striking force in a mobile defense can consist of one-half to two-thirds of the defender's combat power. Division and smaller units generally conduct an area defense or a delay as part of the fixing force as the commander shapes the enemy's penetration, or they attack as part of the striking force. Alternatively, they can constitute a portion of the reserve (FM 3-90).

## SECURITY AREA

3-33. The BCT establishes a security area to provide early warning and reaction time, deny enemy reconnaissance efforts, and protect the MBA. Usually, the BCT executes the forward security mission as a guard or screen. If the division attaches an additional maneuver battalion to the BCT, the battalion may function as a BCT-controlled security force. Typically, there are two options for organizing the security force (Figure 3-1):

- Forward defending combined arms battalions (CAB) or Infantry battalions establish their own security areas.
- CABs or Infantry battalions provide security forces that operate with the reconnaissance squadron under the BCT's direct control.

3-34. The division commander defines the depth of the BCT's security area. The BCT's security area extends from the forward edge of the battle area (FEBA) to the BCT's forward boundary. Depth in the security area provides the MBA forces more reaction time and allows the security force more area to conduct reconnaissance and security tasks and to engage enemy forces. A very shallow security area may require more forces and assets to provide the needed reaction time. The BCT commander must clearly define the objective of the security area. He states the tasks of the security force in terms of time required or expected to maintain security, expected results, disengagement and withdrawal criteria, and follow-on tasks. He identifies specific AAs and named areas of interest (NAI) that the security force must focus on. Security forces also assist the rearward passage of lines of any division and/or corps security forces at the battle handover line (BHL).

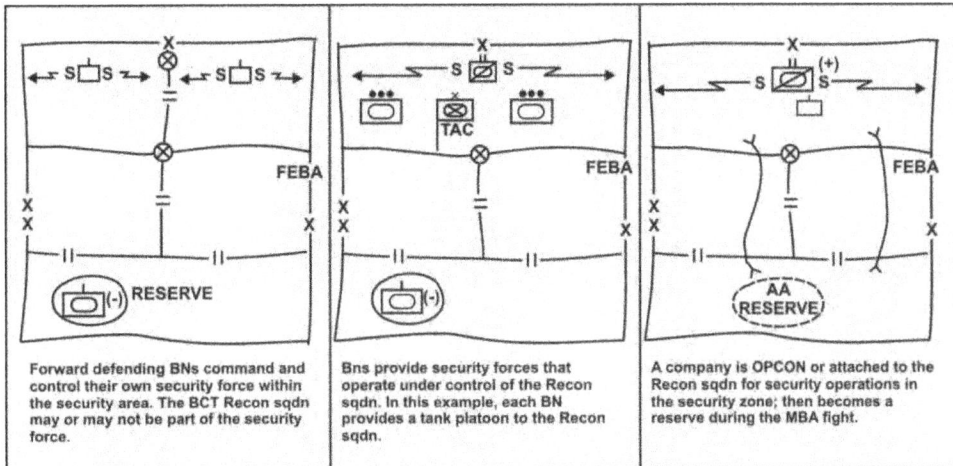

**Figure 3-1. Options for organizing the security area in a contiguous battlefield**

3-35. The BCT conducts security operations outside the battalion AOs but within the outer boundary of the brigade AO. Any maneuver unit (e.g., Infantry, Armor, and reconnaissance), engineer company, or military police (MP) platoon within the BCT can be tasked to conduct security operations in the security area. The BCT gives the unit that conducts security its boundaries to define its area or control measures as part of the overall security plan. Typically, the reconnaissance squadron conducts security operations when threat contact is expected. The MP platoon can conduct security operations when threat contact is not expected.

3-36. Early warnings of pending enemy actions ensure the commander has time to react to any threat. The intelligence staff officer (S-2) analyzes likely routes and methods the enemy could use to conduct reconnaissance. He templates likely locations and activities of enemy observation posts (OP), patrols (mounted and dismounted), and other reconnaissance assets. NAIs are established at these locations to focus counterreconnaissance activities. Security forces use OPs, combat outposts, patrols, sensors, target acquisition radars, and aerial surveillance to locate high-payoff targets (HPT), and to confirm or deny the commander's critical information requirements (CCIR). This is a vital step in disrupting the enemy's plan and getting inside his decision cycle. See Chapter 5 for a detailed discussion of security operations.

## MAIN BATTLE AREA

3-37. The MBA is where the commander intends to defeat the enemy. The BCT's MBA extends from the FEBA to the forward battalion's rear boundary. The commander selects his MBA based on the higher commander's concept of operations, IPB, results of initial reconnaissance and his own assessment of the situation. The commander delegates responsibilities within the MBA by assigning operational areas and establishing boundaries to and for subordinate battalions. If the commander does not assign operational areas to subordinate battalions, the BCT is responsible for terrain management, security, clearance of fires, and coordination of maneuver within the entire AO.

### Area of Operations

3-38. An AO gives maneuver battalions (i.e., CAB or Infantry) freedom of maneuver and fire planning within a specific area. Battalion AOs are situated against enemy brigade AAs. A battalion's AO must provide adequate depth based on its assigned tasks, the terrain, and the anticipated size of the attacking enemy force. To maintain security and a coherent defense, an AO generally requires continuous coordination with flank units. BCT-assigned control measures such as PLs, coordinating points, engagement areas, obstacle belts, and BPs can be used to coordinate battalion defenses within the MBA (Figure 3-2). During defensive preparations, the BCT commander and staff use confirmation briefs, backbriefs, inspections, and supervision to ensure battalion defenses are coordinated, and that unacceptable gaps do not develop. FM 3-90 provides additional information on AOs. Operational areas, battle positions, strongpoints, and combat outposts are all control measures that the commander can use to organize and control his forces.

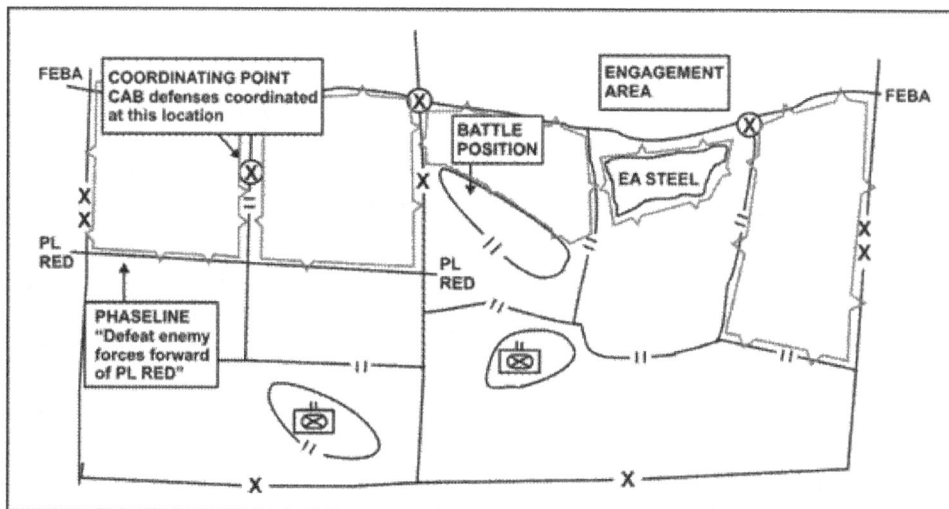

Figure 3-2. Example of control measures used to coordinate defense by area of operation

## Battle Position

3-39. The BCT commander assigns a BP to a battalion when he wishes to control the battalion's fires, maneuver, and positioning. Usually, boundaries are assigned to provide space for the battalion security, support, and sustainment elements that operate outside a BP. When the commander does not establish unit boundaries, the BCT is responsible for fires, security, terrain management, and maneuver between positions of adjacent battalions. The BP prescribes a primary direction of fire by the orientation of the position. The commander defines when and under what conditions the battalion can displace from the BP or maneuver outside it. In addition, the use of prepared or planned BPs, with the associated tasks of prepare or reconnoiter, provides flexibility to rapidly concentrate forces, and adds depth to the defense. Construction of BPs by the BCT requires engineer augmentation.

3-40. When the BCT commander assigns a BP, he considers the following:

- The presence of well-defined enemy brigade-size AAs.
- Selection of terrain that provides sufficient space for dispersion and depth of weapons systems, supplementary and alternate positions, and flanking fires if possible.
- Enemy capabilities.
- Friendly capabilities.

## Strongpoint

3-41. A strongpoint is a heavily fortified BP tied into a natural obstacle or restrictive terrain to create an anchor for the defense. A strongpoint implies retention of terrain for the purpose of controlling key terrain and/or blocking, fixing, or canalizing enemy forces. Strongpoints for armored or mechanized forces are seldom used because they sacrifice the inherent mobility advantage of heavy forces. Before assigning a strongpoint mission, the commander considers the following:

- Loss of survivability and countermobility effort to other areas within the defense.
- Potential for the defending force to be encircled or isolated by the attacking enemy.
- Availability of sufficient time and resources to construct the position.

3-42. Defending units require permission from the higher HQ to withdraw from a strongpoint. Strongpoints are prepared for all-around defense. Strongpoints require extensive engineer effort and resources. All combat, maneuver enhancement, and sustainment assets within the strongpoint require fortified positions. In addition, extensive protective and tactical obstacles are required to provide an all-around defense. A strongpoint usually requires 24 hours of engineer effort by an engineer force equal in size to that of the force defending the strongpoint. Organic BCT engineers lack certain equipment that make the creation of a strongpoint possible within a reasonable amount of time.

## Combat Outpost

3-43. A combat outpost is a reinforced OP that is capable of conducting limited defensive operations. While the factors of METT-TC determine the size, location, and number of combat OPs established by a unit, a reinforced platoon typically occupies a combat outpost. Both mounted and dismounted forces can employ combat outposts. Combat OPs are usually located far enough in front of the protected force to prevent enemy ground reconnaissance elements from observing the actions of the protected force. Considerations for employing combat outposts include:

- They allow security forces to operate in restrictive terrain that prevents mounted security forces from covering the area.
- They can be used when smaller OPs are in danger of being overrun by enemy forces infiltrating the security area.
- They enable a commander to extend the depth of his security area.
- They should not seriously deplete the strength of the main body.

3-44. Forces who man combat OPs can conduct aggressive patrolling, engage and destroy enemy reconnaissance elements, and engage the enemy main body prior to their extraction. The commander should plan to extract his forces from the combat OP before the enemy has the opportunity to overrun them.

## NONCONTIGUOUS AREAS

3-45. The BCT may not have assigned responsibility for all of its AO to its subordinate units. Subordinate unit AOs may be contiguous or noncontiguous. A common boundary separates contiguous areas of operations. Noncontiguous areas of operations do not share a common boundary (Figure 3-3). The concept of operations provides procedural control of elements of the force. The BCT is responsible for controlling the area between noncontiguous areas of operations and/or beyond contiguous areas of operations within its AO.

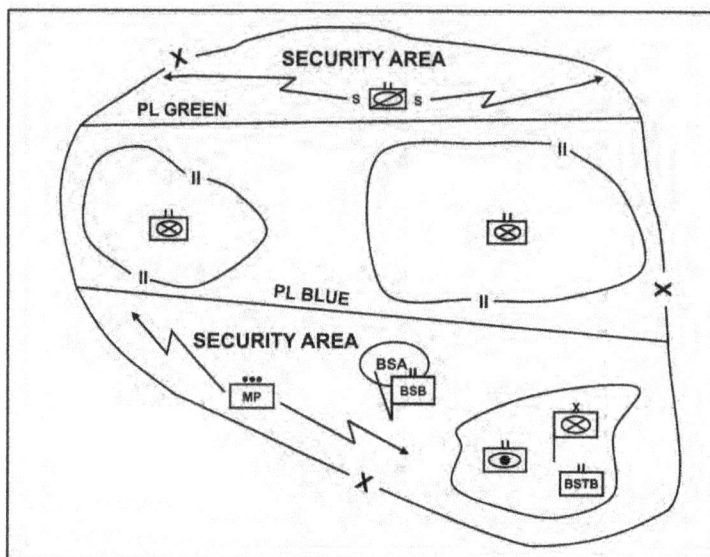

**Figure 3-3. BCT security areas in a noncontiguous battlefield**

## RESERVE

3-46. The reserve is a force(s) withheld from action and committed at a decisive moment. The reserve provides the BCT with the flexibility it needs to exploit success or deal with a tactical setback. The maintenance of a reserve is essential for depth in a defense. The reserve is positioned to respond quickly to unanticipated missions. A reserve maintains protection from enemy fires and detection by maximizing covered and concealed positions, wide dispersion, and frequent repositioning.

3-47. A reserve usually occupies a BP with planning priorities to defend its position; alternately, the reserve can use an assembly area. Maintaining and positioning a reserve is a key requirement for achieving depth within the defense. The commander and staff determine the size and position of the reserve based on the accuracy of knowledge about the enemy and the ability of the terrain to accommodate multiple enemy COAs. When the BCT has good knowledge about the enemy and the enemy's maneuver options are limited, the BCT can maintain a smaller reserve. If knowledge of the enemy is limited and the terrain allows him multiple COAs, then the BCT needs a larger reserve positioned deeper in the AO. This gives the BCT the required combat power and reaction time to commit the reserve effectively.

## OBSTACLES

3-48. The BCT employs tactical obstacles to directly attack the enemy's ability to move, mass, and reinforce. Obstacles are force oriented combat multipliers and usually are covered by observation and direct and/or indirect fires. Tactical obstacles are integrated into the scheme of maneuver and fires to produce specific obstacle effects. Obstacles alone do not produce significant effects against the enemy; obstacles must be integrated with fires to be effective. Fires and obstacles produce four distinct effects: disrupt, fix, turn, and block. There are three types of tactical obstacles: directed, situational, and reserve.

## Directed Obstacles

3-49. Directed obstacles are obstacles assigned by higher commanders as specified tasks to subordinate units. Units plan, prepare, and execute obstacles during the preparation of the defense. The commander can use directed obstacles or obstacle groups to achieve specific obstacle effects at key locations on the battlefield. In this case, the staff plans the obstacle control measures and resources, as well as determines measures and tasks to subordinates to integrate the directed obstacles with fires.

## Situational Obstacles

3-50. Situational obstacles are obstacles the BCT plans and possibly prepares before an operation; however, they do not execute these obstacles unless specific criteria are met. They are be prepared obstacles and provide the commander flexibility for employing tactical obstacles based on battlefield developments. The commander can use engineer forces to emplace tactical obstacles rapidly, but usually uses SCATMINE systems instead. The BCT staff usually plans situational obstacles to enable the commander to shift his countermobility effort rapidly to where he needs it the most, based on the situation. Execution triggers for situational obstacles are integrated into the decision support template. Situational obstacles must be well integrated with tactical plans to avoid fratricide.

## Reserve Obstacles

3-51. Reserve obstacles are obstacles for which the commander restricts execution authority. These are on-order obstacles. The commander specifies the unit(s) responsible for constructing, guarding, and executing the obstacle. Examples of reserve obstacles include preparing a bridge for destruction or an obstacle to close a lane. Units usually prepare reserve obstacles during the preparation phase. They execute the obstacle only on command of the authorizing commander or when specific criteria are met.

## FIRE CONTROL MEASURES

3-52. Chapter 7 discusses the common fire control measures that the BCT commander employs in all types of operations, including the defense.

## SECTION II – PRIMARY DEFENSIVE TASKS

3-53. There are three primary tasks in defensive operations: the area defense, the mobile defense, and the retrograde. These three operations have significantly different concepts and pose significantly different problems (FM 3-90). The BCT commander also must account for the capabilities and limitations of his unit when planning and executing the defense. Although the names of these types of defensive operations convey the overall aim of a selected defensive operation, each typically contains elements of the other and combines static and mobile elements. The BCT's higher headquarters, when it assigns a defensive mission, will also designate the defensive task.

3-54. In an area defense, the BCT concentrates on denying enemy forces access to designated terrain for a specific time limiting their freedom of maneuver and channeling them into killing areas. The BCT retains terrain that the attacker must control in order to advance. The enemy force is drawn into a series of kill zones where it is attacked from mutually supporting positions and destroyed, largely by fires. Most of the defending force is committed to defending positions while the rest is kept in reserve (Figure 3-4). Commanders use the reserve to preserve the integrity of the defense through reinforcement or counterattack (FM 3-0).

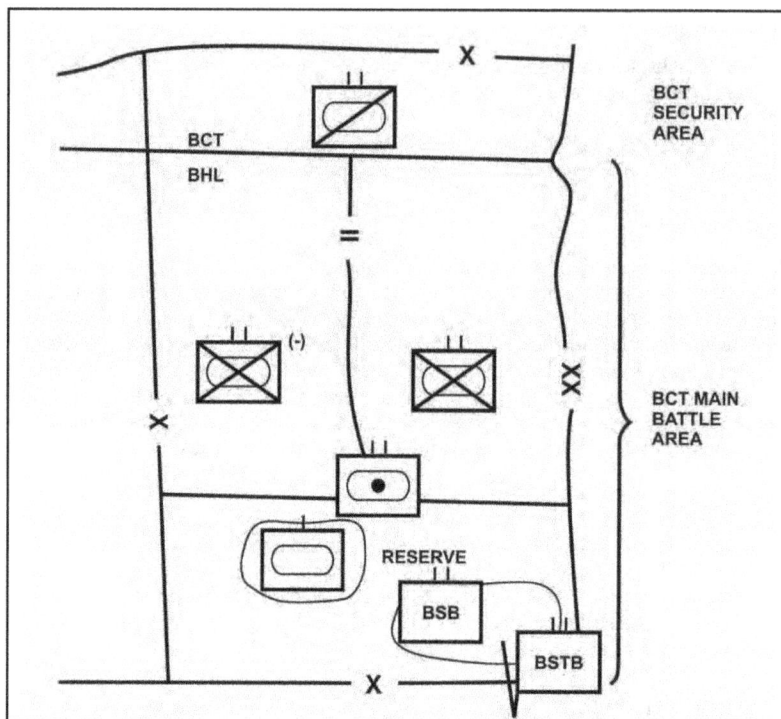

Figure 3-4. Typical HBCT organization of a contiguous area defense

3-55. In a mobile defense, the defender withholds a large portion of available forces for use as a striking force in a counterattack. Mobile defenses require enough depth to let enemy forces advance into a position that exposes them to counterattack. The defense separates attacking forces from their support and disrupts the enemy's C2. As enemy forces extend themselves in the defended area, lose momentum and organization, the defender surprises and overwhelms them with a powerful counterattack (FM 3-0). Divisions and larger forces normally execute mobile defenses. BCTs can participate in a mobile defense as either a fixing force or a striking force (FM 3-90).

# AREA DEFENSE

3-56. An area defense concentrates on denying the enemy's access to designated terrain for a specific time rather than on the outright destruction of the enemy. The keys to a successful area defense are:

- Capability to concentrate effects.
- Depth of the defensive area.
- Security.
- Ability to take full advantage of the terrain such as movement corridors, natural obstacles, and choke points.
- Flexibility of defensive operations.
- Timely resumption of offensive actions.

## SCHEME OF MANEUVER

3-57. The BCT arrays its forces in relationship to likely enemy COAs. The BCT allocates combat forces to the main effort, shaping operation(s), and reserve. Allocations are based on the forces' assigned tasks, the terrain, and the size of enemy force that each avenue of approach can support (probable force ratio). The commander accepts risk along less likely AAs to ensure that adequate combat power is available for more critical efforts. In some cases, the commander must accept gaps within the defense, but must take measures to maintain security within these risk areas. The BCT may use reconnaissance forces, surveillance assets, security forces, patrols, or other economy of force missions for these areas. The BCT uses two forms of defensive maneuver in an area defense: defense in depth and forward defense.

## Defense in Depth

3-58. A defense in depth is the preferred form of maneuver for the Heavy Brigade Combat Team (HBCT) and Stryker Brigade Combat Team (SBCT) because it reduces the risk of the attacking enemy force quickly penetrating the defense. The enemy is unable to exploit a penetration because of additional defensive positions employed in depth. The in depth defense provides more space and time to defeat the enemy attack.

3-59. The BCT uses a defense in depth when:

- The mission allows the BCT to fight throughout the depth of the AO.
- The terrain does not favor a defense well forward, and there is better defensible terrain deeper in the AO.
- Sufficient depth is available in the AO.
- Cover and concealment forward in the AO is limited.
- Weapons of mass destruction might be used.

## Forward Defense

3-60. The intent of a forward defense is to prevent enemy penetration of the defense. Due to its lack of depth, a forward defense is the least preferred form of maneuver. Infantry Brigade Combat Teams (IBCT), which have limited tactical mobility, normally conduct a forward defense. The BCT deploys the majority of its combat power into forward defensive positions near the FEBA (Figure 3-5). The BCT fights to retain its forward position; it may conduct counterattacks against enemy penetrations, or to destroy enemy forces in forward engagement areas. Often, counterattacks are planned forward of the FEBA to defeat the enemy.

3-61. The BCT uses a forward defense when:

- Terrain forward in the AO favors the defense.
- Strong existing natural or manmade obstacles, such as a river or a rail line, are located forward in the AO.
- The assigned AO lacks depth due to the location of the area or facility to be protected. Cover and concealment in the rear portion of the AO is limited.
- Directed by higher HQ to retain or initially control forward terrain.

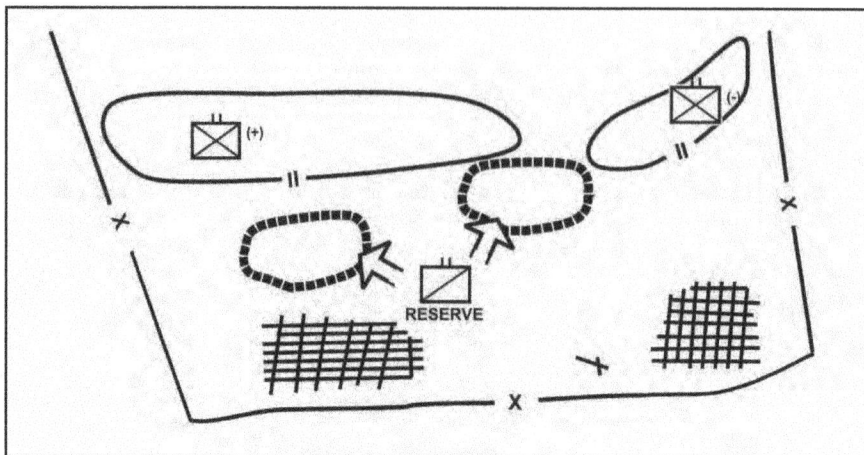

**Figure 3-5. Example of an IBCT forward defense**

### Reconnaissance and Surveillance Considerations

3-62. The BCT commander directs his reconnaissance and surveillance assets to determine the locations, strengths, and probable intentions of the attacking enemy force before and throughout the defensive operation. The commander places a high priority on early identification of the enemy's main effort. He may need to complement surveillance with combat actions that test enemy intentions. Fighting for information can have two benefits—it can force the enemy to reveal his intentions and disrupt his preparations.

3-63. In the defense, reconnaissance operations overlap the unit's planning and preparing phases. Leaders performing reconnaissance tasks must understand that they often deploy before the commander fully develops his plan and they must be responsive to changes in orientation and mission. The commander ensures that his staff fully plans, prepares, and executes reconnaissance missions (FM 3-0).

## PREPARING THE DEFENSE

3-64. The BCT uses the preparation time available to build the strongest defense possible and refining counterattack plans. Commanders and staffs supervise and assess unit preparations while continuing to maintain situational awareness of developments in the AO. Reconnaissance and security operations are conducted aggressively while units occupy their assigned initial positions and rehearse their defensive actions.

### Establish Security

3-65. The first priority in the defense is to establish security. The commander may direct the establishment of a forward security area. He often assigns this mission to the reconnaissance squadron. When defending an extremely wide AO, the reconnaissance squadron establishes a screen. The commander should consider the need to reinforce the reconnaissance and security capabilities of the squadron with tanks and other assets. It is essential that all units maintain a high level of local security. Employment of patrols, establishment of OPs, skillful use of unmanned aircraft systems (UAS) and sensors, and effective use of the terrain to conceal dispositions are essential for effective security.

3-66. In addition to traditional operational security measures, the BCT commander must also consider the potential threat to his defense that non-combatants with access to telephones (landline, wireless, and satellite), digital cameras, and similar devices may present. Security measures, such as shutting down telephone exchanges and cell phone towers, and preventing unauthorized personnel from moving in the defensive area, may be required. The BCT should request guidance from higher HQ before implementing any defensive measures that could affect the civil population.

### Deception and Operational Security

3-67. As part of the defense, higher HQ may have created a deception operation and associated story to protect the force, cause early committal of the enemy, and mislead the enemy as to the true intentions, composition, and disposition of friendly forces. The BCT aids in the execution of the deception plan to:

- Exploit enemy pre-battle force allocation and sustainment decisions.
- Exploit the potential for favorable outcomes of protracted minor engagements and battles.
- Lure the enemy into friendly territory exposing his flanks and rear to attacks.
- Mask the aggressiveness of the sustaining and operational forces committed to the defense.

3-68. Defensive operations contain branches and sequels that give the commander preplanned opportunities to exploit the situation. It is around these branches and sequels that deception potentials exist. Specific deceptive actions the BCT commander can take to hasten exhaustion of the enemy offensive include, but are not limited to:

- Manipulating the SALUTE (size, activity, location, unit, time, and equipment) factors associated with defensive dispositions.
- Masking the conditions under which he will accept decisive battle.
- Luring the enemy into a decisive battle, the outcome of which will facilitate branching or sequencing to the offense.
- Employing camouflage, decoys, false radio traffic, movement of forces, and the digging of false positions and obstacles.

### Occupation of Positions

3-69. As units move into their assigned AOs and occupy positions as directed by the BCT's movement plan, the BCT commander and staff monitor and resolve any problems with the reconnaissance squadron's or higher HQ's reconnaissance and security efforts. The BCT also may have to make minor adjustments to AOs, engagement areas, BPs, and other defensive coordination measures based on unanticipated METT-TC conditions the occupying units encounter as they begin preparing the defense. The BCT should make only critical changes to the defense plan because changes may mean the loss of time and wasted expenditure of unrecoverable CL IV barrier and obstacle material.

3-70. Obstacle emplacement, particularly those directed by higher HQ and the BCT, should be closely monitored to ensure that they are sited and completed in accordance with the obstacle plan. Units assigned to close gaps or to execute engineer targets such as demolition of bridges or dams should also be closely monitored to assure their readiness to execute their mission.

## Rehearsals

3-71. If METT-TC permits, the BCT should rehearse its defense. To prepare for the defense, the commander can use any of the five types of rehearsals. They are confirmation brief, backbrief, combined arms rehearsal, support rehearsal, and battle drill (or standard operating procedure rehearsal). FM 6-0 provides additional information on these rehearsal types. During defensive operations, staffs address counterreconnaissance, battle handover, and then other phases of the operation. The BCT commander also should ensure that his enabling forces are completely integrated into the defensive scheme of maneuver. Rehearsals provide a mechanism for ensuring this integration.

## EXECUTING THE AREA DEFENSE

3-72. In an area defense, the BCT fights mainly from prepared, protected positions to concentrate combat power effects against attempted enemy breakthroughs and flanking movements. The commander uses his reserve to cover gaps between defensive positions, reinforce those positions as necessary, and counterattack to seal penetrations or block enemy attempts at flanking movements.

3-73. The BCT area defense can be described in five steps:

- Gain and maintain enemy contact.
- Disrupt the enemy.

- Fix the enemy.
- Maneuver.
- Follow through.

## Gain and Maintain Enemy Contact

3-74. Gaining and maintaining contact with the enemy in the face of his determined efforts to destroy friendly reconnaissance and security assets is vital to the success of defensive operations. As the enemy's attack begins, the BCT's first concerns are to identify committed enemy units' positions and capabilities, determine the enemy's intent and direction of attack, and gain time to react. Initially, the commander accomplishes these goals in the security area. The sources of this type of intelligence include reconnaissance and security forces, engineer units, intelligence units, special operations forces, and aviation elements. The commander ensures the distribution of a common operational picture (COP) throughout the BCT during the battle to form a shared basis for subordinate commanders' actions. The commander uses the information available to him, in conjunction with his military judgment, to determine the point at which the enemy is committed to a COA (FM 3-90).

### Disrupt the Enemy

3-75. The commander executes his shaping operations to disrupt the enemy regardless of his location within the AO. After making contact with the enemy, the commander seeks to disrupt the enemy's plan, his ability to control his forces, and his combined arms team. Ideally, the results of the commander's shaping operations should result in a disorganized enemy, forced to conduct a movement to contact against prepared defenses. Once the process of disrupting the enemy begins, it continues throughout a defensive operation.

3-76. The BCT uses indirect fires, CAS, attack helicopters, and other available lethal and nonlethal fires during this phase of the battle to:

- Support the security force's delaying action.
- Disrupt or limit the momentum of the enemy's attack.
- Destroy HPTs that support the decisive action of the striking force.
- Divert the enemy's attack.
- Reduce the enemy's combat power.
- Separate enemy formations.

### Fix the Enemy

3-77. The commander has several options to help him fix an attacking force. The commander can design his shaping operations—such as securing the flanks and point of penetration—to fix the enemy and allow friendly forces to execute decisive maneuver elsewhere. As discussed earlier in this chapter, combat outposts and strong points can also deny enemy movement to or through a given location. A properly executed military deception operation can constrain the enemy to a given COA.

3-78. The commander limits the options available to the enemy by using obstacles covered by fire to fix, turn, block, or disrupt enemy activities. Properly executed obstacles are a result of the synthesis of top-down and bottom-up obstacle planning and emplacement. Blocking forces can also affect enemy movement. A blocking force can achieve its mission from a variety of positions depending on the factors of METT-TC.

### Maneuver

3-79. In an area defense, the decisive operation occurs in the MBA. This is where the effects of shaping operations, coupled with sustaining operations, combine with the decisive operations of the MBA force to defeat the enemy. The commander's goal is to prevent the enemy's further advance through a combination of fires from prepared positions, obstacles, and mobile reserves.

3-80. Generating massed effects is especially critical to the commander conducting the defense of a large area against an enemy with a significant advantage in combat power. The attacker has the ability to select the point and time of the attack. Therefore, the attacking enemy can mass his forces at a specific point, dramatically influencing the ratio of forces at the point of attack. An enemy three-to-one advantage in overall combat power can easily turn into a local six-to-one or higher ratio. The defending commander must quickly determine the intent of the enemy commander and the effects of terrain. This allows his units and their weapons systems to use agility and flexibility to generate the effects of combat power against the enemy at those points, and restore a more favorable force ratio.

*Follow Through*

3-81. The purpose of defensive operations is to retain terrain and create conditions for a counteroffensive that regains the initiative. The area defense does this by causing the enemy to sustain unacceptable losses short of his decisive objectives. A successful area defense allows the commander to transition to an attack. An area defense could also result in a stalemate with both forces left in contact with each other. Finally, it could result in the defender being overcome by the enemy attack and needing to transition to a retrograde operation. Any decision to withdraw must take into account the current situation in adjacent defensive areas. Only the commander who ordered the defense can designate a new FEBA or authorize a retrograde operation.

3-82. During this follow-through period, time is critical. Unless the commander has a large, uncommitted reserve prepared to quickly exploit or reverse the situation, he must reset his defense as well as maintain contact with the enemy. Time is also critical to the enemy, because he will use it to reorganize, establish a security area, and fortify his positions.

## Security Area Engagement

3-83. When the tasks of the division's security force mission are met, and the security force begins its movement from the security area to the flanks of the division or rearward, the security force passes a final enemy spot report (SPOTREP) to the BCT (via FM voice and digital), and the BCT assumes control of the battle.

3-84. BCT security forces observe and maintain contact with the approaching enemy, report enemy movements, avoid decisive engagement, and withdraw as lead enemy formations enter the BCT's security area. The commander can direct security forces to disrupt, delay, or destroy lead portions of the enemy formations. The commander also can include his security forces as part of an effort to deceive the enemy as to the actual location of the MBA. The commander must consider the follow-on missions of his security forces, the potential for these forces to be over-run or isolated, and the overall impact their direct combat achieves. To remain abreast of the situation and maintain mutual support, main battle forces eavesdrop on the security force fight.

3-85. The BCT continues to disrupt the tempo of the approaching enemy formations to ensure these forces are unable to restore any lost momentum. The commander usually maintains indirect fires and CAS on the approaching enemy formations as they enter the MBA. The commander must clearly state his essential tasks for fire support and essential tasks for mobility-survivability for this phase of the defense. In addition, he must ensure his fire support systems remain responsive in order to weight their efforts to the MBA fight as it develops. The commander orders the execution of situational obstacles that best support the MBA engagement; controls occupation of defensive positions; and assesses the impact of fires against the enemy. At this point, the commander's major concerns are to identify the enemy's main effort, determine the direction of the attack, and gain time to react.

*Battle Handover*

3-86. The battle handover (BHO) is the transfer of responsibility for the battle from BCT security forces to the BCT combat forces in the MBA. The BCT commander prescribes criteria for the handover. He designates where forces will pass through, routes, contact points, and the BHL. The BHL is usually forward of the FEBA, which is where elements of the passing unit are effectively over-watched by direct fires of the forward combat elements. Battalions usually employ security forces in the area immediately behind the BHL. Each maneuver battalion coordinates the BHO with the security force to their front. This coordination overlaps the coordination for the passage of lines, and so the battalion should conduct the two simultaneously. To facilitate a rapid BHO, it is best to establish its coordination as standard operating procedure (SOP). BHO coordination usually includes:

- Establishing communications.
- Providing updates on both friendly and enemy situations.
- Coordinating passage.
- Collocating C2.
- Dispatching representatives to contact points and establishing liaison.
- Recognition signals.
- Status of obstacles and routes.
- Fire support, air defense, and sustainment requirements.
- Defining exact locations of contact points, lanes, and other control measures.
- Assisting the security force when breaking enemy contact.
- Coordinating and exchanging maneuver, obstacle, and fire plans.
- ROE.
- Civilian considerations (to include displaced persons).

3-87. While a line defines the BHO, events might force the security force to break contact forward of, or behind the BHL, or in the gaps that develop between attacking enemy echelons. As security force elements cross the BHL, each battalion directs its movements along designated routes through their battalion AO. As needed, battalion security forces and fire support systems assist the passing force to break contact with the enemy. Mass artillery, obscurants, CAS, and possibly situational obstacles are used to support the break in contact. Commanders must closely coordinate control of fires to avoid fratricide and ensure effective fire distribution during execution. Close coordination at all levels is essential to execute this process. The BHO is completed when the passing unit is clear and the battalions have assumed control of the battle.

3-88. The entire security force should not withdraw automatically as soon as the first enemy units reach the FEBA. The commander can leave in place security elements located in areas where the enemy has not advanced. A single force in the security area can perform both reconnaissance and security functions. The security force uses every opportunity for limited offensive action to delay and harass the enemy and to gain information (FM 3-90).

3-89. The security force adjusts to the enemy's advance and continues to conduct security operations as far forward as possible. It continues to resist the enemy's shaping operations, such as the enemy's reconnaissance effort, thereby upsetting his coordination and enabling the MBA commander to fight one engagement (or battle) at a time. Doing this increases the chances for success even if the enemy attack penetrates into the MBA in some areas. In some cases, the security force can attack the enemy force from its rear, engage HPTs, or drive between echelons to isolate leading enemy units.

3-90. During BHO, each maneuver battalion in the MBA:

- Assists passage of lines and disengagement.
- Gains and maintains contact with enemy forces as the BHO occurs.
- Continues to locate and destroy enemy reconnaissance and security elements to preclude observation of the primary defensive positions.
- Closes lanes, executes reserve obstacles, and/or emplaces situational obstacles in the security area as the passing force withdraws.

## MAIN BATTLE AREA ENGAGEMENT

3-91. All systems and units focus on fixing and destroying enemy forces that enter the MBA. Only through decisive combat can the BCT defeat/destroy a determined enemy and complete its mission.

## Maneuver

3-92. During the MBA engagement, the BCT shifts combat power and priority of fires to defeat the enemy's attack. This may require:

- Adjusting subordinates' AOs and missions.
- Repositioning forces.
- Shifting the main effort.
- Repeating commitment and reconstitution of a reserve.
- Modifying the original plan.

3-93. Forward forces within the MBA usually have the mission to break the enemy's momentum, reduce his numerical advantage, and force the enemy into positions of vulnerability. The BCT masses combat power at decisive times and locations to counter major enemy efforts and defeat enemy formations. The BCT economizes and takes risks in less threatened areas, shifts fires, and maneuvers the reserve and/or MBA forces to gain local fire superiority at critical locations. Obstacles, security forces, surveillance assets, and fires can assist covering areas where risk is accepted. Often, the BCT must trade ground to gain the time needed to concentrate forces, mass fires, and attrit the enemy. The commander must decide, and must mass forces and fires swiftly, since periods that allow him to gain an advantage usually are brief.

## Maintain Cohesion

3-94. The BCT must maintain a cohesive defense if the defense is to remain viable. The commander ensures battalion movements do not uncover adjacent battalions or adjacent BCTs. Often, the BCT must accept gaps in the defense. In such cases, the commander must take measures to cover these gaps and detect enemy efforts towards these risk areas. The BCT commander and subordinate commanders use security forces, surveillance assets, and patrols to maintain a cohesive defense.

3-95. Subordinate commanders cross talk and continually report their situation, enemy actions, and future plans to the BCT commander. Subordinate commanders accomplish this via the digitally displayed COP. The commander assesses individual battalion plans to ensure they are consistent with his scheme of maneuver. Often, defending battalions must modify their defensive plans to protect and refuse their flanks when necessary actions of an adjacent maneuver battalion/CAB create an assailable flank. The BCT commander must ensure all battalion actions are coordinated and controlled to provide a cohesive defense.

## Penetrations

3-96. Each battalion commander is responsible for controlling enemy advances within his assigned AO. Battalion commanders must provide the BCT commander early warning and reaction time for potential enemy penetrations. If a battalion is threatened with a penetration the BCT commander may take several actions to counter the situation. In order of priority, he can do any or all of the following:

- Maintain contact with the penetrating enemy force. Forward MBA forces may be able to transition into a delay to maintain contact, or the commander may redirect reconnaissance assets, security forces, and observers to locate and observe the enemy. The commander seeks to determine the penetrating enemy force's size, composition, direction of attack, and rate of movement. Forces in contact also must adjust indirect fires and CAS against the enemy to disrupt, delay, or divert his attack.
- Take immediate actions to hold the shoulders of the penetration. This may require changing task organization, adjusting adjacent maneuver battalion/CAB boundaries and tasks, executing situational or reserve obstacles, or shifting priority of fires.
- Move threatened sustainment units. Based on the enemy's direction of attack, sustainment units may need to move away from the penetration. These movements must be controlled to ensure they do not interfere with counterattack plans or movements of combat forces.
- Determine where and how to engage the penetrating enemy force. Based on the enemy's size, composition, and direction of attack the commander selects the best location to engage the enemy. The reserve may counterattack into the enemy's flank, or it may establish a defensive position in depth to defeat or block the enemy. The staff establishes control measures for the

reserve's attack. The reserve can use an engagement area or objective to orient itself to a specific location to engage the enemy. The reserve can use a BP as a position along defensible terrain. When the situation is vague or the enemy has multiple AAs, the commander may establish an AO for the reserve. This requires the reserve to locate, and move to intercept and engage the enemy anywhere in the assigned AO. The commander and staff develop a concept of fires and consider required adjustments to fire support coordinating measures. They also decide on the commitment of directed, reserve, or situational obstacles to support the action. Traffic control is especially critical. Sufficient routes must be designated for the reserve to use, and provisions such as the use of MPs must be taken to ensure those routes remain clear.

- Issue an order. If the operation is not well controlled, the situation could easily deteriorate into a total force failure. The BCT commander must develop orders quickly, and issue them clearly, concisely, and calmly. A simple, well thought-out plan, developed during the initial planning process, greatly improves the ability of subordinates to react effectively.

3-97. The BCT commander must keep the division commander well informed of the BCT's situation. Potential enemy penetrations of the BCT's AO are immediately addressed to the division commander. Depending on the resources available, the division commander might reinforce the BCT with additional fires, CAS, attack aviation, or maneuver forces.

### Counterattack to an Enemy Penetration

3-98. The BCT conducts counterattacks to take advantage of an attacking enemy's weakened condition by striking against his flanks or rear, or to deny the enemy commander momentum and initiative. As the enemy's advance slows and weakens, he has fewer maneuver options. As a result, he could transition to a hasty defense along the forward line of own troops (FLOT), or he could attempt to gain a foothold within the brigade's MBA from which he can defend.

3-99. This situation enables the BCT commander to seek decisive opportunities to counterattack the enemy with all available force, and ultimately secure the initiative of the battle. Timing is critical to a counterattack. Assuring the mobility of the counterattack force is critical. If committed too soon, the counterattack force might not have the desired effect, or may not be available for a more dangerous contingency. If committed too late, they might be ineffective. Once committed, counterattack forces can penetrate the enemy's flanks and attack the enemy's artillery and logistics areas; or penetrate the enemy's flanks and attack them from the rear. Both of these are decisive actions, and create concern for the enemy.

3-100. After a successful counterattack, and once minimum reorganization activities are completed, the commander orders his forces to attack key enemy objectives in order to place the BCT in positions for future operations. As the BCT reaches the objective of the attack, it consolidates and continues reorganization that is more extensive. It then begins preparation to resume future offensive operations.

## MOBILE DEFENSE

3-101. A mobile defense is a force oriented defensive action that focuses on the destruction of the enemy rather than the retention of terrain. Terrain is traded to overextend the attacker and diminish his ability to react to counterattacks. A mobile defense requires considerable depth. Divisions, Corps, or larger formations normally conduct a mobile defense.

3-102. BCTs participating in a mobile defense usually function as a security force, a striking force, or as part of a fixing force. The fixing force holds attacking enemy forces in position to help channel attacking enemy forces into ambush areas, and to retain areas from which to launch the striking force. The striking force is a dedicated counterattack force constituting the bulk of available combat power. The decisive operation is a counterattack conducted by the striking force.

3-103. IBCTs usually are not used in the mobile defense due to their limited mobility. The SBCT can be either the fixing or the striking force, but their lack of offensive Armor generally limits their employment as part of the fixing force. Within a division mobile defense, HBCTs are usually the striking force, though they are fully capable of being the fixing force. See FM 3-90 for additional information on mobile defenses.

# RETROGRADE OPERATIONS

3-104. Retrograde operations are defensive, organized movements that direct troops away from an enemy (Table 3-1). Forces use retrograde operations to protect an overwhelmed or weakened force, or to improve an untenable tactical situation (FM 3-90). In either case, the BCT's higher HQ must approve the operation. Retrograde operations accomplish the following:

- Resist, exhaust, and damage enemy forces, while avoiding becoming decisively engaged.
- Draw the enemy into an unfavorable situation.
- Gain time.
- Preserve combat power.
- Disengage from battle for use elsewhere in other missions.
- Reposition forces or shorten lines of communication (LOC).

### Table 3-1. Forms of retrograde operations

| Operation | Intent | Threat Contact |
|-----------|--------|----------------|
| Delay | Trade space for time. | In contact. |
| Withdrawal | Disengage force free unit for use elsewhere. | In contact. |
| Retirement | Move a force away from the threat. | Not in contact. |

3-105. An integral part of successful retrograde operations is disciplined execution. Movement to the rear may be seen as defeat or threat of isolation unless Soldiers have confidence in their leaders, and understand the purpose of the operation and their role in it. Leaders must be present, display confidence in the plan, be in control of the battlefield, and thoroughly brief Soldiers on their role in the overall operation. Leaders must remind Soldiers that they are conducting combat operations that will free the unit for other operations, while continuing to inflict casualties upon the enemy.

3-106. BCTs must preserve their freedom to maneuver. While a portion of the unit may become decisively engaged, the commander cannot allow the entire unit to do so. He must develop contingencies and be prepared to fight to free battalions or companies that can no longer extricate themselves.

## DELAY

3-107. In a delay, the BCT trades space for time and inflicts maximum damage on the enemy while attempting to avoid decisive engagement. Usually, inflicting damage is secondary to gaining time. The BCT may execute a delay when it has insufficient combat power to attack or defend or when the higher unit's plan calls for drawing the enemy into an engagement area or area for a counterattack. Delays gain time to:

- Allow other friendly forces to establish a defense.
- Cover a withdrawing force.
- Function as an economy of force effort to enable other forces to counterattack.

3-108. The two types of delay missions are:

- Delay within an AO. This mission is used to slow and defeat as much of the enemy as possible without sacrificing the tactical integrity of the unit. It presents low risk to the unit.
- Delay forward of a specific area or position for a specific period of time. This mission is used to slow an enemy advance for a specific period of time, or defeat specified enemy formations within an area to support the higher commander's concept of operations. This can involve engagement with all or part of the unit, and presents high risk to the unit.

### Organization

3-109. The BCT usually organizes into a security force, main body, and reserve. The main body consists of the majority of the BCT's combat power, and usually deploys well forward within the AO. The security force usually establishes a screen forward of the initial positions of the main body. The security force then may be tasked to become the reserve when it is withdrawn from the security area. The reserve may have a mission to contain, or defeat enemy penetrations between delay positions, conduct limited objective counterattacks, or assist other units to break contact. Sustainment assets tend to be widely dispersed and often attached to the units they support.

### Force the Enemy to Deploy and Maneuver

3-110.   Commanders should select terrain that supports engagements at maximum weapons ranges. This causes the enemy to take time-consuming measures to deploy, develop the situation, and maneuver to drive the delaying force from its position. The delay must be sufficiently tenacious to make the enemy doubt the nature of the friendly mission and leave it no choice but to deploy and maneuver.

### Maintain a Mobility Advantage Over the Attacker

3-111.   Maintaining a mobility advantage over the attacking enemy is essential to a successful delay. The goal is to increase the BCT's mobility while degrading the enemy's ability to move. The BCT improves its mobility by task organizing appropriate engineer assets within subordinate formations, using and rehearsing multiple routes, displacing nonessential sustainment elements early in the operation, and having a rapid casualty evacuation (CASEVAC) plan. Mobility operations facilitate this aspect of the scheme of maneuver and enable the BCT to achieve this.

3-112.   The BCT degrades the mobility of the enemy by destroying enemy disruption and security forces, controlling dominant terrain, engaging at maximum ranges, extensive use of obstacles, and synchronized effects of lethal and nonlethal effects.

### Alternate and Subsequent Positions

3-113.   In planning, the commander chooses to delay from either alternate positions or subsequent positions. In a delay from alternate positions, two or more units in a single AO occupy delaying positions in depth (Figure 3-6). As the first unit engages the enemy, the second occupies the next position in depth and prepares to assume responsibility for the operation. The first force disengages and passes around or through the second force. It then moves to the next position and prepares to reengage the enemy while the second force takes up the fight.

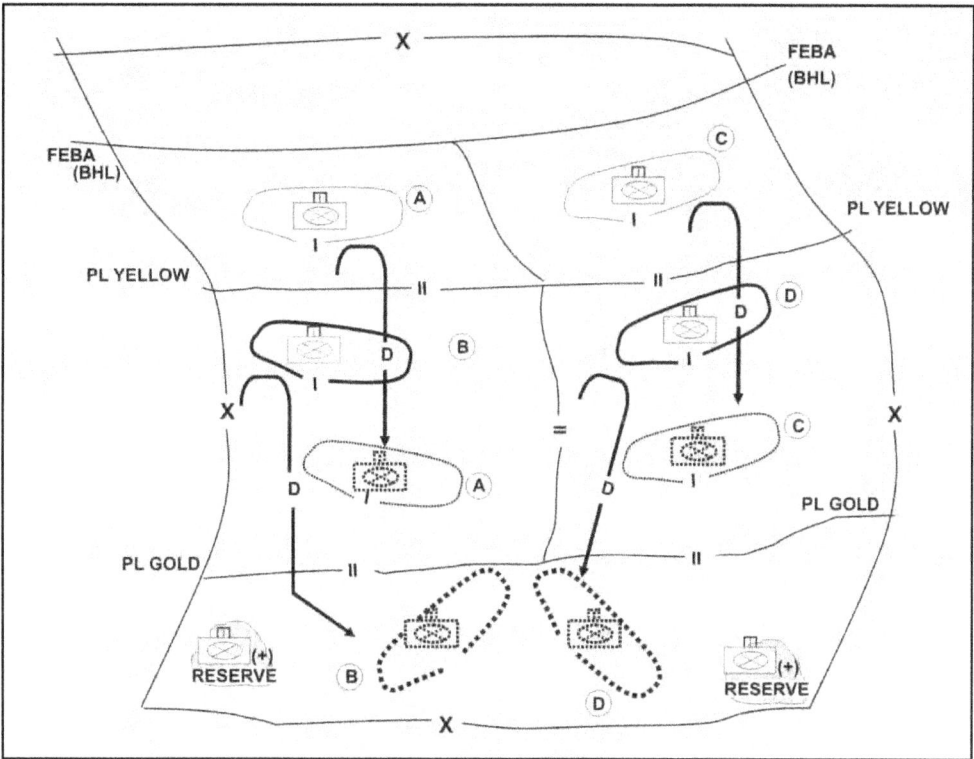

**Figure 3-6. Delay by alternate positions**

3-114. The BCT uses a delay from subsequent positions when the assigned AO is so wide that available forces cannot occupy more than a single tier of positions across a front (Figure 3-7). In a delay from subsequent positions, the majority of forces are arrayed along the same PL or series of BPs. There are no forces in depth; only unoccupied positions. The forward forces delay the enemy from one PL, reposition to the next PL, and then the same forces delay the enemy again.

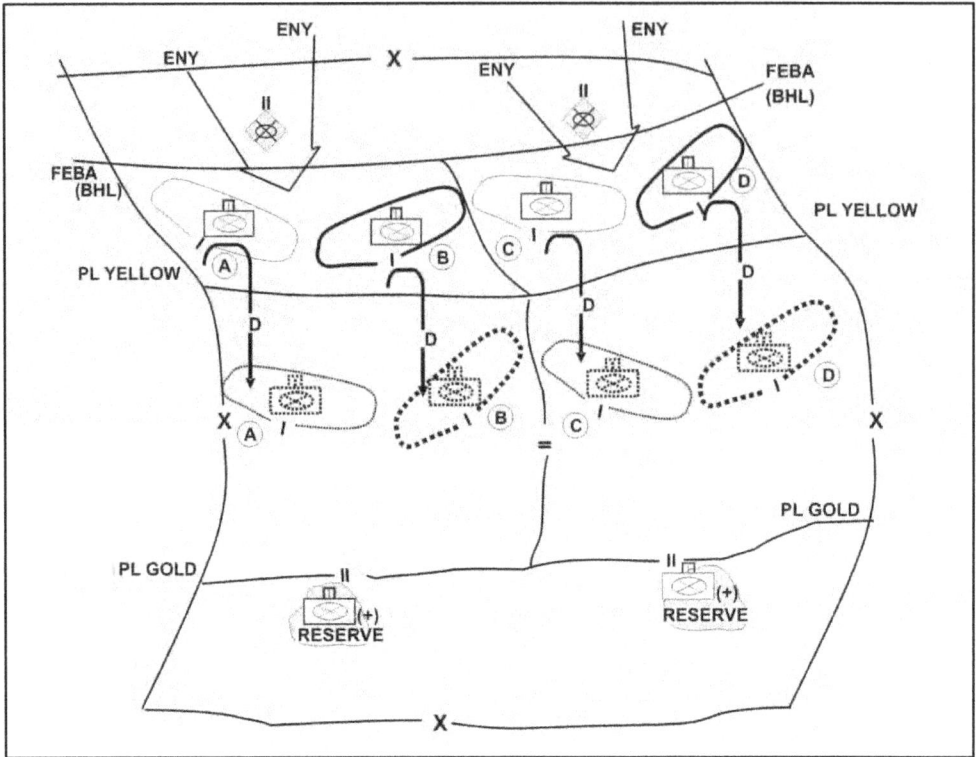

Figure 3-7. Delay by subsequent positions

3-115.  In both techniques, the delaying forces maintain contact with the enemy between delay positions. Table 3-2 details the advantages and disadvantages of the two techniques.

Table 3-2. Comparison of methods of delay

| Method Of Delay | Use When . . . | Advantages | Disadvantages |
|---|---|---|---|
| Delay from subsequent positions. | AO is wide. Forces available do not allow themselves to be split. | Masses fires of all available combat elements. | Limited depth to the delay positions. Less time is available to prepare each position. Less flexibility. |
| Delay from alternate positions. | AO is narrow. Forces are large enough to be split between different positions. | Allows positioning in depth. Allows more time for equipment and Soldier maintenance. More flexibility. | Requires continuous coordination. Requires passage of lines. Only part of the force is engaged at one time. |

**Controlling the Delay**

3-116.  The commander usually decentralizes execution of the delay to maneuver battalion/CAB level. He must rely on his subordinate commanders to execute their mission and request help if and when they need it. Subordinates displace once they meet previously established parameters. Units delaying within the BCT must continuously coordinate movement and actions with units surrounding them. Displacements may be preplanned events or time dependent. The commander closely controls the disposition of his forces to maintain cohesion and control of the situation.

3-117. The BCT commander directs or allows delays from one position or PL to the next only after considering the following:

- What are the strengths, compositions, and dispositions of attacking enemy forces? Are elements of the BCT threatened with decisive engagement or bypass?
- What is the status of adjacent units? How does their status affect the BCT's capability to continue to delay?
- Does the movement affect the cohesion of the BCT's disposition? Are other movements necessary to maintain cohesion? Do any sustainment assets need to reposition?
- What is the condition of the delaying force in terms of troops, equipment, and morale?
- How strong is this position in relation to other positions that may be occupied?
- Is unit survivability or time key to the mission and higher commander's intent?

## Counterattacks

3-118. Whenever possible, the BCT takes any opportunity to seize the initiative, even if only temporarily. By aggressively contesting the enemy's initiative through offensive action, the BCT avoids passive defensive patterns that favor the attacking enemy. Counterattacks disorganize the enemy, confuse the enemy commander's picture of the situation, and help prolong the delay. Counterattacks also affect the enemy's momentum.

## Decisive Engagement

3-119. Friendly forces usually do not become decisively engaged. There are two exceptions to this. First, when engagement is necessary to prevent the enemy from prematurely advancing across a piece of key terrain. Second, when a part of the force must be risked to prevent jeopardizing the integrity of the whole force.

3-120. If elements of the BCT are threatened with decisive engagement or have become decisively engaged, the commander must take actions to support their disengagement. In order of priority, he can do any of the following:

- Allocate priority of all supporting fires to the threatened unit. This is the most rapid and responsive means of increasing the unit's combat power.
- Reinforce the unit.
- Conduct a counterattack to disengage the unit.

## Terminating the Delay

3-121. A delay mission ends with another planned mission such as a defense, withdrawal, or attack. If the enemy reaches his culmination point during the delay, the BCT can maintain contact while another force counterattacks, withdraw to perform another mission, or transition to the offense. In all cases, the commander must plan for the expected outcome of the delay based on the situation and the higher commander's plan.

## WITHDRAWAL

3-122. The withdrawal is a planned operation in which a force in contact disengages from an enemy force. Withdrawals may or may not be conducted under enemy pressure (Figure 3-8). The two types of withdrawals are:

- **Assisted.** The assisting force occupies positions to the rear of the withdrawing unit and prepares to accept control of the situation. In addition, it can assist the withdrawing unit with route reconnaissance, route maintenance, fire support, and sustainment. Both forces coordinate the withdrawal closely. Once plans are coordinated, the withdrawing unit delays to a BHL, conducts a passage of lines, and moves to its final destination.

- **Unassisted.** The withdrawing unit establishes routes and develops plans for the withdrawal, then establishes a security force as the rear guard while the main body withdraws. Sustainment elements usually withdraw first followed by combat forces. The BCT may establish a detachment left in contact (DLIC) if withdrawing under enemy pressure. The DLIC's goal is to deceive the enemy as to the friendly movement. As the BCT withdraws, the DLIC disengages from the enemy and follows the main body to its final destination.

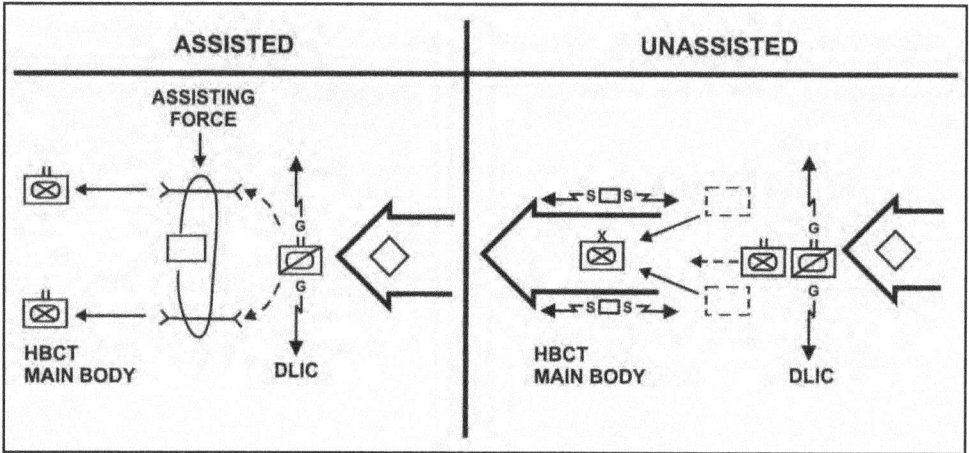

**Figure 3-8. Types of withdrawals**

### Organization

3-123. The BCT usually organizes into a security force, main body, and reserve. It can also organize a DLIC or stay behind forces if required by the enemy situation. Usually, the reconnaissance squadron is the security force; the maneuver battalions/CABs are the main body; and a maneuver company is the reserve.

3-124. The security force maintains contact with the enemy until ordered to disengage, or until another force takes over the task. It simulates the continued presence of the main body. This requires allocation of additional combat multipliers to a reconnaissance unit. When the BCT conducts withdrawal without enemy pressure, the security force transitions into a rear guard because the most probable threat is a pursuing enemy. When withdrawing under enemy pressure, the security force establishes or operates as a detachment left in contact to provide a way to sequentially break contact with the enemy.

3-125. A DLIC is an element that is left in contact as part of the previously designated (usually rear) security force, while the main body conducts its withdrawal. Its purpose is to remain behind to deceive the enemy into believing the BCT is still in position while the majority of the BCT withdraws. The commander must establish specific instructions about what to do if the enemy attacks and when and under what circumstances to delay or withdraw. The BCT organizes a DLIC in one of three ways (Figure 3-9):

- The reconnaissance squadron or a single maneuver battalion/CAB operates as the DLIC.
- Each maneuver battalion/CAB provides forces for the DLIC mission, which then operates under the BCT's control.
- Each maneuver battalion/CAB establishes and controls their individual DLIC.

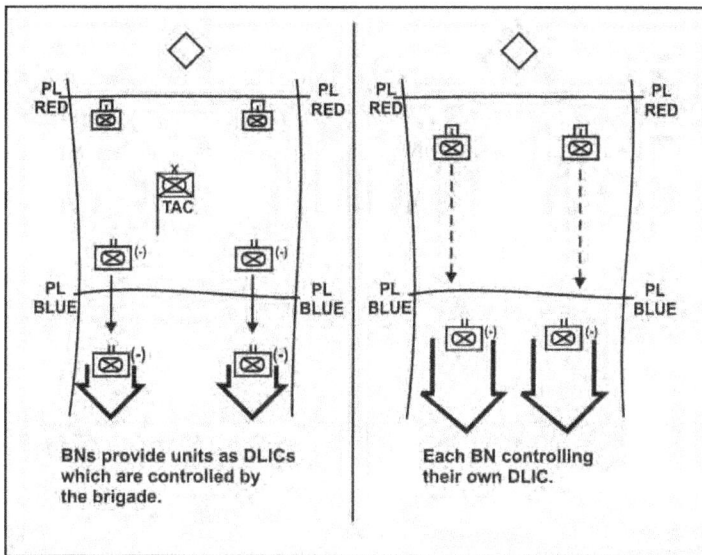

Figure 3-9. Method for organizing the detachment left in contact

3-126. The main body consists of all elements except the security force and reserve. The main body withdraws along pre-designated routes to its final destination. The main body maintains all-around security during the withdrawal and movement.

3-127. The reserve provides the BCT with the flexibility to deal with unexpected enemy actions. The reserve may take limited offensive action such as spoiling attacks to disorganize or disrupt the enemy. It can counter enemy attacks, reinforce threatened areas, and protect withdrawal routes.

3-128. The commander develops his vision of the battle based on withdrawing under enemy pressure. He must determine the composition and strength of the security force, main body, and reserve. The commander must clearly define how he intends to deceive the enemy by executing a withdrawal; how he intends to disengage from the enemy (use of maneuver, fires, and obstacles); and the end state of the operation in terms of time, location, and disposition of forces.

## Disengagement

3-129. The security force remains in position and maintains a deception while the main body moves as rapidly as possible rearward to intermediate or final positions. After the main body withdraws a safe distance, the commander orders the security force to begin its rearward movement. Once the security force begins moving, it assumes the duties of a rear guard. If the enemy is not pursuing the BCT, the security force can move in a march column.

3-130. The main body moves rapidly on multiple routes to designated positions. It can occupy a series of intermediate positions before completing the withdrawal. Usually sustainment units, along with their convoy escorts, move first and precede combat units in the withdrawal movement formation. The staff enforces the disciplined use of routes during the withdrawal. Despite confusion and enemy pressure, subordinate units must follow specified routes and movement times.

## Terminating the Withdrawal

3-131. Once the BCT successfully disengages from the enemy, it usually has the following options:
- Rejoin the overall defense under favorable conditions.
- Transition into a retirement.

- Continue moving away from the enemy and towards its next mission.
- The higher HQ define the BCT's next mission. Follow-on missions usually are planned as the BCT is preparing for, or executing, the withdrawal.

## RETIREMENT

3-132.  A retirement is a retrograde operation in which a force that is not in contact with the enemy moves to the rear in an organized manner. The BCT usually conducts a retirement to reposition for future operations. The BCT usually organizes into security elements and a main body. The formation and number of columns employed depend on the number of available routes and the potential for enemy interference. The commander typically wants to move his major elements to the rear simultaneously. During a retirement, the BCT usually moves to an assembly area to prepare for future operations. The elements of the BCT move in accordance with established movement times and routes. Strict adherence to the movement plan is essential to avoid congestion. The staff closely supervises the execution of the movement plan. Sustainment and assets (units) not attached or organic to the maneuver battalions/CABs or separate companies usually move to the rear first, followed by combat forces.

## SECTION III – TRANSITIONS

3-133.  During the planning for any operation, the BCT commander and staff must discern from the higher HQ operations order (OPORD) what the potential follow-on missions are and begin to plan how they intend to achieve them. Whether the BCT is concluding an offensive or defensive operation, it must pause to consolidate and reorganize before the next operation. If required, the commander decides the best time and location that facilitates future operations and provides protection. The BCT must maintain a high degree of security when performing consolidation and reorganization activities.

# CONSOLIDATION

3-134.  Consolidation is the process of organizing and strengthening a newly occupied position. The BCT might need to consolidate in order to reorganize, avoid culmination, prepare for an enemy counterattack, or allow time for movement of adjacent units. Consolidation is planned for every mission. Actions during consolidation include:

- Maintain contact with the enemy and conduct reconnaissance.
- Establish security consistent with the threat.
- Eliminate pockets of enemy resistance.
- Position forces to enable them to conduct a hasty defense by blocking possible enemy counterattacks.
- Clear obstacles or improve lanes to support friendly movement and reorganization activities.
- Plan and prepare for future operations.

3-135.  The BCT maintains contact with the enemy by redirecting reconnaissance and surveillance assets, directing small-unit patrols, and possibly conducting limited objective attacks. In some situations, the BCT might leave a small force to control key terrain or complete clearing the objective while the remainder of the BCT transitions to a new mission.

# REORGANIZATION

3-136.  Reorganization refers to all measures taken to maintain the combat effectiveness of the BCT or return it to a specified level of combat capability. All units undertake reorganization activities during operations, as the situation allows, to maintain combat effectiveness. More extensive reorganization is usually conducted after the BCT defeats an enemy attack. Reorganization tasks usually include:

- Establish security consistent with the threat. This may include moving forces, adjusting boundaries, changing task organization, and adjacent unit coordination.
- Destroy or contain enemy forces that still threaten the BCT.
- Replace or shift reconnaissance and surveillance assets, if needed.

- Reestablish the BCT chain of command, key staff positions, and C2 facilities lost during the battle.
- Treat and evacuate casualties.
- Redistribute ammunition, supplies, and equipment as necessary.
- Conduct emergency resupply and refueling operations.
- Recover and repair damaged equipment.
- Send relevant logistics and battle reports by digital means and voice (if not digitally equipped).
- Process enemy prisoners of war (EPW) and detainees as required.
- Repair/emplace additional obstacles and improve/construct additional fighting positions.
- Repair/restore critical routes within the BCT AO to assure mobility of the force.
- Reposition C2 facilities, communications assets, logistics, and fire support assets for future operations.

# CONTINUING OPERATIONS

3-137. At the conclusion of an engagement, the BCT may continue the defense, or if ordered, transition to offensive or stability operations. The BCT commander considers the higher commander's concept of operations, friendly capabilities, and the enemy situation when making this decision. All missions should include plans for exploiting success or assuming a defense.

## OFFENSE

3-138. Higher headquarters may order the BCT to conduct a hasty attack, movement to contact, or participate in exploitation. In some cases, the defensive operation might immediately transition into a pursuit. If reorganization is required, the BCT maintains pressure on the enemy through artillery, CAS, and/or limited objective attacks.

## STABILITY OPERATIONS

3-139. BCT commanders must ensure that transitions from defensive operations to stability operations and vice versa are planned. For example, it may be tactically wise for commanders to plan a defensive contingency with on-order offensive missions for certain stability operations that could deteriorate. Subordinate commanders and leaders must be fully trained to recognize activities that would initiate this transition. Commanders, staffs, and Soldiers must be aware that elements of the BCT could be conducting offensive, defensive, and stability operations simultaneously within a small radius of each other. Actions in one unit's AO can affect a change in whatever type operation an adjacent unit is conducting. For example, an offensive operation may result in displacing noncombatants to another section of the city, thus creating stability operations for the unit in that AO.

This page intentionally left blank.

# Chapter 4

# Stability Operations

Stability operations are a fundamental aspect of Brigade Combat Team (BCT) full spectrum operations. Stability operations focus activity on maintaining or reestablishing a safe and secure environment, and on providing essential services, emergency infrastructure reconstruction, and humanitarian relief. They lead to an end state that, in support of a legitimate government, enables these activities to facilitate other instruments of national power. Stability operations can be conducted simultaneously with offensive or defensive operations, tasks, and activities in which the needs of the population must be addressed immediately. In some situations, stability operations may be the predominant activity in the BCT area of operations (AO). The primary characteristic of stability operations is the focus on nonlethal support, in conjunction with non-military organizations, for the civilian population backed up by lethal and nonlethal security activity.

## SECTION I – OVERVIEW

4-1. Stability operations encompass various military missions, tasks, and activities conducted outside the United States in coordination with other instruments of national power to maintain or reestablish a safe and secure environment, provide essential governmental services, emergency infrastructure reconstruction, and humanitarian relief. Stability operations:

- Leverage the coercive and constructive capabilities of the BCT to establish a safe and secure environment.
- Facilitate reconciliation among local or regional adversaries.
- Establish political, legal, social, and economic institutions within the BCT AO.
- Facilitate the transition of responsibility to a legitimate civilian authority.

## RESPONSIBILITIES

4-2. The primary responsibility for providing basic civil functions rests with the host nation government or civilian organizations. When this is not possible, the BCT establishes basic civil functions within its area of operations and protects them until a civil authority can provide these services for the local populace. The BCT performs specific functions in support of a broader effort by other government and nongovernment organizations.

## TASKS

4-3. The stability operations framework helps the BCT determine the required training and task organization of forces prior to initial deployment, and serves as a guide to action in stability operations (FM 3-07). This framework encompasses the tasks that the BCT expects to perform during stability operations. The tasks performed during stability operations may be framed as:

- **Initial response tasks.** These tasks generally reflect activity executed to stabilize the area of operations in a crisis state. The BCT typically performs initial response tasks during, or directly after, a conflict or disaster in which the security situation prohibits the introduction of civilian personnel. Initial response tasks aim to provide a secure environment that allows relief forces to attend to the immediate humanitarian needs of the local population. They reduce the level of violence and human suffering while creating conditions that enable other actors to participate safely in relief efforts. Examples of other actors are joint interagency, intergovernmental, multinational (JIIM) agencies, nongovernmental organizations, and contractors.

- **Transformation tasks.** These tasks represent the broad range of stabilization, reconstruction, and capacity-building performed in a relatively secure environment. Transformation phase tasks take place in either crisis or vulnerable states. These tasks aim to build host-nation capacity across multiple sectors. While establishing conditions that facilitate broad unified action to rebuild the host nation and its supporting institutions, these tasks are essential to ensuring the continued stability of the environment.

- **Fostering sustainability tasks.** These are tasks that encompass long-term efforts, which capitalize on capacity building and reconstruction activities. Successful accomplishment of these tasks establishes conditions that enable sustainable development. Usually military forces perform fostering sustainability phase tasks only when the security environment is stable enough to support efforts to implement the long-term programs that commit to the viability of the institutions and economy of the host nation. Often military forces conduct these long-term efforts to support broader, civilian-led efforts.

4-4. Stability operations consist of five primary tasks the BCT may be assigned to perform or support:

- Establish civil security.
- Establish civil control.
- Restore essential services.
- Provide support to governance.
- Provide support to economic and infrastructure development.

## SECTION II – BRIGADE COMBAT TEAM STABILITY TASKS

4-5. Success in stability operations depends on the commander's ability to identify the tasks essential to mission success from the wide range of possible stability tasks. Success also depends on prioritizing and sequencing the execution of those tasks with available combat power, the diverse array of actors participating, and the ability of the population in the AO to accept change. Stability operations require commanders to demonstrate a knowledge of the local culture to determine which stability tasks are essential to mission success. Essential stability tasks may not become clear until the BCT has occupied the area and performed reconnaissance to identify local requirements.

4-6. The BCT may be directed to conduct any of the stability tasks, but may require augmentation to perform them. Additionally, a description of the full list of primary stability tasks is provided since BCTs may be required by their headquarters to execute these tasks as well. For more detail on stability tasks, see FM 3-07.

# ESTABLISH CIVIL SECURITY

4-7. Establishing a safe, secure, and stable environment for the local populace within the BCT area of operations is a key to obtaining their support for the overall operation. Such an environment allows the introduction of the civilian agencies and organizations whose efforts ensure long-term success. When the people have confidence in the security sector providing for their safety, they offer the cooperation required to control crime and subversive behavior, defeat insurgents, and limit the effects of adversaries. For political and economic reform to be successful, people, commodities, and currency must be able to freely flow throughout the region.

4-8. The BCT can conduct civil security tasks without significant augmentation. These tasks are characterized by the use of area security, site exploitation operations, civil-military operations, and information engagement tactics, techniques, and procedures (TTP) as described in other areas of this manual.

4-9. Subcategory tasks the BCT might be directed to perform include:
- Enforce cessation of hostilities, peace agreements, and other arrangements.
  - Enforce cease fires.
  - Supervise disengagement of belligerent forces.
  - Identify and neutralize potential adversaries.
  - Provide security for negotiations.
  - Protect and secure strategically important institutions (such as government buildings; medical and public health infrastructure; the central bank, national treasury, and integral commercial banks; museums; and religious sites).
  - Protect and secure military depots, equipment, ammunition dumps, and means of communications.
  - Identify, secure, protect, and coordinate the disposition of stockpiles of munitions and chemical, biological, radiological, and nuclear (CBRN) materiel and precursors; facilities; and adversaries with technical expertise.
  - Build local host nation capacity to protect civilian reconstruction and stabilization personnel.
  - Build host nation capacity to protect infrastructure and public institutions.
  - Build host nation capacity to protect military infrastructure.
  - Protect host nation high risk personnel necessary for governance and security.
- Clear explosive and CBRN hazards.
  - Conduct emergency clearing of mines, unexploded ordnance (UXO), and other explosive hazards.
  - Map, survey, and mark mined areas, UXO, and other explosive hazards.
  - Remediate hazards remaining from the release of CBRN hazards and radiological fallout, as well as provide decontamination support.
  - Create local host nation capacity to conduct de-mining.
  - Initiate a deliberate process to warn and alert the local population to dangers of UXO.

*Note.* The John Warner National Defense Authorization Act limits the assistance that military forces may provide with respect to de-mining. The BCT may assist and train others in de-mining techniques and procedures. However, no member of the armed forces—while providing humanitarian de-mining assistance—will engage in the physical detection, lifting, or destroying of landmines or other explosive remnants of war. The exception to this rule is if the member does so for the concurrent purpose of supporting a U.S. military operation, such as route or area clearance. Nor will any member provide such assistance as part of a military operation that does not involve the armed forces.

# ESTABLISH CIVIL CONTROL

4-10. Civil control regulates selected behavior and activities of individuals and groups. It reduces risk to individuals or groups, and promotes security. Initial response tasks aim to develop interim enforcement mechanisms for establishing rule of law. These tasks typically involve assessing and building indigenous police, and penal capacity and capability. Transformation tasks focus on restoring the justice system and processes for reconciliation. Fostering sustainability tasks serve to establish a legitimate, functioning justice system founded on international norms. These conditions define success within the AO while reflecting the end state needed to ensure the foundation for enduring stability and peace. The BCT should expect substantial augmentation by civil affairs and military police elements to conduct this primary stability task properly.

4-11. Subcategory tasks the BCT may be directed to perform include:
- Establish public order and safety.
- Establish interim criminal justice system.

- Support law enforcement and police reform.
- Support judicial reform.
- Support property dispute resolution processes.
- Support legal system reform.
- Support human rights initiatives.
- Support corrections reform.
- Support war crimes courts and tribunals.
- Support public outreach and community rebuilding programs.
- Determine disposition and constitution of national armed and intelligence services.
  - Implement a plan for disposition of local security forces, intelligence services, and other local security institutions.
  - Identify future roles, missions, and structure for multinational forces in the BCT AO.
  - Evaluate senior officers and other individuals for past abuses and criminal activity.
  - Conduct security force assistance (includes training of host nation military and police forces.
  - Build local host nation capacity to protect military infrastructure in the AO.
  - Establish security institutions in the AO.
  - Support military training and cooperation programs with host nation forces and services.
- Disarmament, demobilization, and reintegration.
  - Negotiate arrangements with belligerents using leader engagement tasks. See Chapter 8 for a discussion of leader engagement tasks.
  - Establish and enforce weapons control programs including collection and destruction.
  - Provide reassurances and incentives for disarmed factions.
  - Establish a monitoring program.
  - Establish internment camps for demobilization.
  - Ensure adequate health, food, and security for belligerents.
  - Disarm former combatants and belligerents.
  - Reduce availability of unauthorized weapons.
  - Ensure safety of quartered personnel and families.
  - Secure, store, and dispose of weapons.
  - Develop host nation arms control capacity.
  - Reintegrate former combatants and dislocated civilians into society.
  - Identify and separate extremists.
- Conduct border control, boundary security, and freedom of movement.
  - Establish border control and boundary security.
  - Establish, disseminate, and enforce rules relevant to movement.
  - Dismantle roadblocks and establish checkpoints.
  - Ensure freedom of movement.
  - Train and equip border control and boundary security forces.
  - Position sustainment areas, bases, and facilities for protection and rapid movement.
- Support identification.
  - Secure documents relating to personal identification, property ownership, court records, voter registries, professional certificates, birth records, and driving licenses. See FM 3-90.15 for information on site exploitation.
  - Establish an identification program.
  - Ensure individuals have personal forms of identification.
  - Support the development of mechanisms for long-term dispute resolution over property rights and family or group affiliations within the AO.
  - Ensure that host nation authorities are credentialed or uniformed as needed.

- Protect key personnel and facilities.
  - Protect government-sponsored civilian reconstruction and stabilization personnel.
  - Protect contractor and civilian reconstruction and stabilization personnel and resources in the AO.
  - Provide emergency logistics support, as required.
  - Protect and secure places of religious worship and cultural sites.
  - Protect and secure critical infrastructure, natural resources, civil registries, and property ownership documents.

# RESTORE ESSENTIAL SERVICES

4-12. The activities associated with this primary stability task extend beyond simply restoring local civil services and addressing the effects of humanitarian crises. While the BCT generally centers efforts on the initial response tasks for immediate needs of the populace, other civilian agencies and organizations focus on broader humanitarian issues. Normally, the BCT supports host nation and civilian relief agencies with these efforts. However, when the host nation cannot perform its roles, the BCT may execute these tasks directly. The BCT need not wait for large-scale projects that require complicated national-level efforts to sustain them when the necessary infrastructure is not yet in place to support such an effort. The BCT and other actors should begin ground level restoration efforts as soon as requirements in their AO surface that can be addressed with the resources available and within the constraints of the current security environment. The performance of this primary stability task is characterized by substantial interaction and cooperation with a wide range of JIIM, nongovernmental organizations (NGO), and contractor elements. The BCT engineer coordinator (ENCOORD) can coordinate the capabilities of the U.S. Army Corps of Engineers (USACE) field force engineering to help in the restoration of essential services. Subcategory tasks the BCT may be directed to perform include:

- Provide essential civil services.
- Perform tasks related to civilian dislocation.
- Support famine prevention and emergency food relief programs.
- Support shelter and relief programs other than food relief programs.
- Support humanitarian de-mining.
- Support public health programs.
- Support education programs.

# SUPPORT TO GOVERNANCE

4-13. When a legitimate and functional local government is present in the AO, the BCT operating in support of local governance has a limited role. However, if the local government cannot adequately perform its basic civil functions—for whatever the reason—some degree of military support to governance may be necessary. A government's legitimacy among its people is tied in part to its perceived ability to provide these essential services. In extreme cases, the civil government may be completely dysfunctional or absent altogether. In such cases, international law requires the military force to provide the basic civil administration functions of the host nation government under the auspices of a transitional military authority. FM 3-07 describes various aspects of transitional military authority, including its legal basis, command responsibilities, organizing principles, guidelines, the role of courts, and claim processes. Performing this primary stability task requires extensive civil affairs support.

4-14. Subcategory tasks the BCT may be directed to perform include:

- Support transitional administrations.
- Support development of local governance.
- Support anticorruption initiatives.
- Support elections.

# SUPPORT TO ECONOMIC AND INFRASTRUCTURE DEVELOPMENT

4-15. The BCT owning an AO may be required to support local economic and infrastructure development. This can require augmentation by civil affairs and financial management elements, and is characterized by extensive interaction with JIIM organizations, NGOs, and contractors. Subcategory tasks the BCT may be directed to perform include:

- Support economic generation.
- Support monetary institutions and programs.
- Support national treasury operations.
- Support public sector investment programs.
- Support private sector development.
- Protect natural resources and environment.
- Support agricultural development programs.
- Restore transportation infrastructure.
- Restore telecommunications infrastructure.
- Support general infrastructure reconstruction programs.

4-16. BCT-funded jobs programs are a common method to stimulate the local economy and to reduce causes of local violence. Developing the public sector is usually managed by organizations such as the United States Department of State (DOS) and the United States Agency for International Development. Developing the private sector typically begins with employing large portions of the labor force. The agricultural sector is a cornerstone of a viable market economy providing crops and livestock vital to local markets. FM 3-07 gives a full discussion of the activities that may be performed or supported by the BCT if the commander determines that these stability tasks are required.

## SECTION III – CONSIDERATIONS FOR STABILITY OPERATIONS

4-17. The expeditionary capabilities of the BCT enable it to move promptly into any operational environment. In an operational environment with unstable security conditions, where the host nation government is unable to function effectively, the BCT may be the only substantial stabilizing presence. In these conditions, the BCT is organized and prepared to perform all the tasks essential to establishing and maintaining civil security and civil control, while providing the essential needs of the populace. In many situations, local and international aid organizations will be present in the operational area but may have limited access to the population. The BCT can significantly contribute to increasing the access of these aid organizations, enabling them to provide essential humanitarian assistance to the civilian population. In turn, this reduces a substantial logistic burden on the BCT, allowing it to focus on providing a safe, secure environment.

4-18. Stability operations support the complete range of joint military operations, operational themes, and the full spectrum of conflict. Many types of operations (e.g., counterinsurgency) rely extensively on stability operations due to their focus on the population. However, the BCT may execute stability operations throughout the spectrum of conflict whenever it is too dangerous for less armed and protected forces to do so. This implies a level of combat readiness inherent to all stability activities against threats ranging from civilian lawlessness to highly organized insurgents.

4-19. All stability tasks such as policing, intelligence, civil-military operations (CMO) and trash collection involve working with other U.S. government agencies and host nation partners. Soldier training for these operations is enhanced by briefings from the DOS, aid agencies, and the local police or fire departments. Units use cultural advisors from those who come from that nation or culture to get advice concerning local issues. Military operations create the environment for civilian agencies to achieve needed long-term development and stabilization.

# LETHAL AND NONLETHAL ACTIONS

4-20. During full spectrum operations, a complementary relationship exists between the BCT's lethal and nonlethal actions. Every situation requires a different combination of violence and restraint. Nonlethal actions are vital contributors to all operations but may be decisive in the execution of stability tasks. Determining the appropriate combination of lethal and nonlethal actions necessary to accomplish the mission is an important consideration for every commander. Nonlethal actions expand the options available to commanders to achieve their objectives. Conditions may limit the conduct of lethal actions, and forces must be organized appropriately to reflect this change in emphasis. At tactical levels, the line between lethal and nonlethal activities can be blurred and expressed as rules of engagement or security force assistance. Most stability tasks conducted by military forces have elements of both lethal and nonlethal capabilities, and mission planning should consider this expectation.

4-21. BCTs have a potent combination of lethal and nonlethal capabilities. The mere presence of military forces often influences human behavior, as demonstrating the potential for lethal action helps to maintain order. Maintaining order is vital to establishing a safe, secure environment. Even though stability operations emphasize nonlethal actions, the ability to engage potential adversaries with decisive lethal force remains a sound deterrent and is often a key to success. The successful application of lethal capabilities in stability operations requires a thorough understanding of when the escalation of force is necessary and when it might be counterproductive. Rules of engagement and rules on the use of deadly force should be continuously assessed to match conditions in the AO and reflect the commander's intent for Soldier and unit actions.

4-22. Public perception is a major consideration for the BCT. Every action in the public eye has consequences, intended and often unintended. These consequences may be positive or negative to achieving the commander's end state. Planning takes these potential outcomes into account as much as possible. The actions of Soldiers influence how the local populace perceives the military. Leaders manage local perceptions, inform the populace about friendly intentions, and may need to explain their actions. This is accomplished through information engagement, which is detailed in Chapter 8, Section VI of this manual. Commanders use information engagement by leaders and Soldiers to inform, influence, and persuade the populace within limits prescribed by international law. In this way, commanders enhance the legitimacy of the operation and the credibility of friendly forces. Stability operations are conducted within several constraints that BCT leaders must be aware of. The BCT operational law team compiles a list and summary of rules, agreements, and other constraints relevant to the mission. The list below is not comprehensive, but gives an example of the sources that may be referenced by BCT leaders in planning and training stability tasks:

- Codes of conduct.
- The Hague and Geneva conventions.
- United Nation mandates.
- Status of forces agreements.
- Multi-national agreements.
- Contractor agreements.

4-23. One planning approach is to identify phases of the operation in terms of major objectives to achieve along lines of effort such as establishing dominance, building local networks, and marginalizing the enemy. Forces should easily transition between phases: forward to exploit successes, and backward to recover from setbacks. Insurgents adapt their activities to friendly tactics, so the plan must be simple enough to survive setbacks without collapsing.

# WARFIGHTING FUNCTION CONSIDERATIONS

4-24. As the BCT commander develops his commander's intent and concept of operations, he sets priorities for each warfighting function. The following information describes some of the considerations the BCT commander uses to set those priorities.

## MOVEMENT AND MANEUVER

4-25. If another unit is performing the mission before the BCT occupies its AO, a formal transfer of authority or mission reassignment occurs as directed by the higher headquarters or coordinated between the BCT commander and his counterpart. The incoming BCT assumes tactical responsibility for its assigned base(s) and the AO at that time. Prior to transfer of authority or assumption of the mission, the unit to be replaced retains command of the operations and may control the movements of the incoming BCT units as provided for by the order. Once transfer of authority or mission change has occurred, the incoming BCT exercises tactical control of departing elements for security and mission performance. Command relationships must be established and coordinated before the incoming BCT begins to occupy its AO.

### Establish Presence

4-26. Establishing the force's presence in the AO is often the first requirement of stability operations. This can require living in the AO close to the populace. Being on the ground establishes links with the local populace. Through Soldier engagement, the populace begins to trust and relate to friendly forces. Driving around in an armored convoy may degrade situational awareness (SA). It can make Soldiers targets and is often more dangerous than moving on foot and remaining close to the populace.

4-27. Upon arrival in the AO, it may not be advisable to go straight for the main insurgent stronghold or to try to take on villages that support insurgents. Start from secure areas and work gradually outwards. Extend influence through local networks. First, win the confidence of a few villages, and then work with those with whom they trade, intermarry, or do business. This tactic develops local allies, a mobilized populace, and trusted networks.

4-28. Seek a victory early in the operation to demonstrate dominance of the AO. This does not require a combat victory. Early combat without accurate situational understanding may create unnecessary collateral damage and ill will. Instead, victories may involve using leader engagement to resolve a long-standing issue or co-opt a key local leader. Achieving even a small early victory can set the tone for the mission and help commanders seize the initiative.

### Create Conditions for Small Unit Success

4-29. Commanders work stability problems collectively with subordinate leaders who own the ground in their respective areas, sharing understanding and exploring possible solutions. Once leaders understand the situation, seeking consensus helps subordinates understand the commander's intent. Subordinates need to exercise initiative and act based on the commander's intent informed by whatever situational awareness they have developed. Employing mission command is essential in this environment.

4-30. The brigade maintains the ability to conduct coordinated small-scale operations over great distances quickly and securely. Due to the multiple and unique demands of these operations, supporting forces such as engineers, logistics, and medical personnel must remain responsive and flexible. Task organization of augmenting units will often change many times during the course of operations. The brigade must ensure adequate support for its subordinate units and take active measures to create the conditions for its subordinates to succeed.

### Employ Quick Reaction Forces

4-31. The BCT normally establishes one or more quick reaction forces for the security of its checkpoints, outposts, observation posts, and work sites, and to support patrols, meetings, and convoys in the AO. Planning should also provide forces of appropriate size for a quick reaction force to separate local hostile parties before potential violent situations grow out of control. The force must have the ability to respond anywhere in the brigade area and be rapidly reinforced by augmentation and maneuver elements.

### Modify Tactics as Conditions Evolve

4-32. External events can sometimes negate local advances achieved in stability operations. The BCT commander and staff assess, consolidate, regain situational understanding, and prepare the BCT to expand control and security again when the situation allows. A flexible, adaptive plan helps in such situations.

Friendly forces may have to cede the initiative for a time; however, they must regain it as soon as the situation allows.

4-33. Refine the plan implemented early in the operation through interaction with local partners. BCTs should aim at dominating the whole AO and implementing solutions to systemic problems. BCTs continuously assess results and adjust as needed. Achieving success means that, particularly late in the operation, it may be necessary to negotiate with the enemy. Local people supporting stability operations know the enemy's leaders. Valid negotiating partners sometimes emerge as the operation progresses. Use close interagency relationships to exploit opportunities to co-opt segments of the enemy with leader engagement and stability task activity. This helps counteract the insurgency without alienating potential local allies who have relatives or friends among insurgents.

4-34. More planning considerations for the movement and maneuver warfighting function can be found in Chapter 5. Stability operations can also include offensive and defensive tasks, and the continuing activity of reconnaissance and security operations.

## INTELLIGENCE

4-35. Once the BCT occupies the AO, its next task is to build trusted networks. Over time, successful trusted networks grow like roots into the populace. They displace enemy networks, which forces enemies into the open, letting military forces seize the initiative and destroy the insurgents. Trusted networks are diverse, including local allies, community leaders, and local security forces. Networks should also include NGOs, other friendly or neutral nonstate actors in the AO, and the media. Building trusted networks begins with conducting village and neighborhood surveys across the BCT AO to identify community needs. Then follow through to meet them, build common interests, and mobilize popular support. This becomes the main effort in stability operations. Actions that help build trusted networks support the stability effort. Actions that undermine trust or disrupt these networks, even those that provide a short-term military advantage, help the enemy.

4-36. Stability operations rely on an extensive understanding (i.e., past and current estimates) of immediate and adjacent operational environments, including the people, demographics, infrastructure, topography, economy, history, religion, and culture. Leaders must be aware of every village, road, field, population group, tribal leader, and ancient grievance. If the precise area of operations has not yet been assigned, commanders and staff study the general area until the precise destination is determined. During premobilization and mobilization training, and reset/train Army forces generation (ARFORGEN) phases, ensure leaders and staffs use the Secret Internet Protocol Router Network (SIPR) to immerse themselves virtually in the AO into which the unit is deploying. Conducting simulations on actual terrain maps can greatly enhance knowledge and understanding of the AO and potential enemy activity in it. Understand factors in adjacent areas and the information environment that can influence the BCT AO.

4-37. Gain an understanding and seek insights of what motivates the local people and apply that information and knowledge to mobilize them. A clear and detailed understanding of why and how insurgents attract and recruit followers is critical to understanding the tactical circumstances. Insurgents are adaptive, resourceful, and probably from the area. The local populace has known them since they were young. The Soldiers of the BCT are the outsiders. Insurgents are not necessarily misled or naive; much of their success may stem from bad government policies or host nation security forces that alienates the local populace.

4-38. Threat mitigation during stability operations is intelligence driven, and units often develop much of their own intelligence in relation to the amount they receive from higher headquarters (HQ). BCT commanders organize their assets to collect local information unavailable to higher sources of intelligence. Augmentation for extra intelligence positions is normally not available, but staffs still must perform tasks such as human mapping and local intelligence collection. Appendix A of FM 3-24.2 contains a detailed explanation of the relevance of intelligence preparation of the battlefield (IPB) in counterinsurgency (COIN).

4-39. Linguists are a key asset, but like any other scarce resource, commanders must allocate them carefully. During predeployment, the best use of linguists may be to train BCT Soldiers in basic language skills.

### Civil Reconnaissance

4-40. An additional form of reconnaissance now recognized by the Army, and one that would be critical to the successful execution of BCT stability operations (and civil support operations) is civil reconnaissance (CR). CR is defined in FM 1-02 as A targeted, planned, and coordinated observation and evaluation of those specific civil aspects of the environment. Civil reconnaissance focuses specifically on the civil component, the elements of which are best represented by the mnemonic ASCOPE: areas, structures, capabilities, organizations, people, and events. Civil reconnaissance can be conducted by civil affairs personnel or by other forces, as required." It differs from other reconnaissance in that it usually is not targeted at a specific enemy; instead, it focuses on answering information requirements for civil situation awareness.

4-41. When a mission analysis identifies information requirements (IR) that require CR, then a CR plan is prepared in support of civil information management (CIM) that will provide the necessary situational awareness for the civil component of the BCT's common operating picture (COP). Most commonly, a CR plan will be prepared by the BCT S-9, possibly in conjunction with a nonlethal working group, and with the support of a civil affairs (CA) company. The integrator of the information, when attached to the BCT, is the civil-military operations center (CMOC) section of the supporting CA company. However, the CR plan itself is a BCT operation and must be coordinated through the S-3's current operations and plans sections. While CA units may be used as the primary collectors in a CR plan, other BCT organic and attached elements may also be tasked to collect information. In addition, civil reconnaissance can by coordinated and synchronized as a joint effort with JIIM elements.

4-42. Because civil situational awareness represents the terrain (i.e., the "platform") on which stability operations are implemented, it is important that CIM reporting be structured in a manner useful for commanders and subordinate units. One effective method is the application of the civil considerations of ASCOPE to each of the operational variables of politics, military, economic, social, information, and infrastructure, plus physical environment and time (PMESII-PT) of a selected area. The S-2 and S-3 usually cover the military component. Further information on civil reconnaissance is available in FM 3-05.40 (under revision) and FM 3-05.401. Another source is FM 3-24.2, Chapter 1, Sect. III (pages 1-8 through 1-16).

4-43. Appendix B in FM 3-24 contains a discussion of social network analysis and other analytical tools. It can be useful for promoting situational understanding of the operational environment for stability operations as well as COIN. FM 3-07, Appendix D, describes the State Department's Interagency Conflict Assessment Framework, a tool for assessing conflict situations systemically and collaboratively. It supports U.S. Government interagency planning for conflict prevention, mitigation, and stabilization.

## Fires

4-44. In stability operations, artillery units can perform such doctrinal roles as:
- Quick reaction fire support for patrols and counterinsurgency operations.
- Mortar and rocket counterfire.
- Show of force fire missions.
- Base security.

4-45. Artillery, and air and missile defense units and staff must also be prepared to execute tasks such as:
- Information engagement.
- Civil military operations.
- Local, area, route, and convoy security.
- Host nation security force training.

4-46. In some situations, the BCT may find it advantageous to develop fire bases across its AO to position quick fire support for dangerous areas. For mutual security, these should be collocated with forward operating bases of maneuver forces.

4-47. Other planning considerations for fire support include:

- Developing procedures for the rapid clearance of fires.
- Increasing local security for firing positions of indirect weapons.
- Enabling 360 degree firing capability in positions.
- Coordinating with host nation officials and security forces in areas of operations.
- Establishing communications with host nation forces and area control centers.
- Understanding the restrictions on the use of dual-purpose improved conventional munitions and area denial antipersonnel mine/remote antiarmor mine system.
- Using illumination rounds to defuse belligerent's night activities.
- Using radars/artillery in a protection role for rapid targeting and suppression of indirect fire attacks.

## SUSTAINMENT

4-48. The capability of the brigade to sustain itself is a function of the theater's maturity, the sustainment structure, and the flow of forces into the AO. Sustainment for stability operations is unique and more complex due to physically dispersed unit locations, lack of adequate infrastructure, nontraditional demands by civil military operations, and the burden caused by displaced civilians. Planning considerations for conducting sustainment in this type of environment include:

- Flexibility to support varying task organizations.
- Indigenous support through the use of contracting and local purchase of supplies, facilities, utilities, services, labor/manpower, and transportation support systems.
- Existing indigenous facilities such as roads, ports, airfields, and communications systems.
- Development or improvement of the indigenous capabilities for self-support for the eventual transfer of responsibilities to the host nation.
- Economy of resources.
- Availability and employment of health services.
- Sustainment elements may provide support for coalition, governmental agencies, and civilians when authorized by law.
- Operational contract support is an effective force multiplier.
- Increased consumption of classes I, III, IV, and VIII supplies.
- Requirement to sustain internment/resettlement compounds, facilities, and camps.

4-49. Sustainment assets require hardening and reliable communications to reduce their vulnerability to attack. Prepare Soldiers for local security requirements whose primary task is providing logistics support.

4-50. There is an emphasis on sustainment by air when the roads are unsecure in the AO. Maintain a close working relationship between the BCT and brigade support battalion (BSB) S-4 staffs, and supporting Army and Air Force air transportation elements.

4-51. BCT sustainment assets and their augmentations sustain not only the military forces, but also may provide support for civil-military operations. Some considerations include:

- In an AO with a low priority for resources, commanders focus on self-reliance, keeping work projects small and sustainable, and prioritizing efforts.
- Local leaders are helpful in prioritizing, as they know what matters most to them. Commanders should be honest with them, discuss possible projects and options, and ask them to recommend priorities.
- Often commanders can find translators, building supplies, or expertise in the local area.
- Locals might only require protection in completing their projects.
- Negotiation and consultation can help mobilize their support and strengthen social cohesion.
- Setting achievable goals is a key to making the situation work.

## PROTECTION

4-52. Protection of the force during stability operations is essential for success at all levels. Commanders continually balance protection needs between military forces and civil populations. Frequent interaction between U.S. forces and the local population make protection planning difficult and essential. Adversaries often blend in with the local populace during stability operations and are difficult to identify, making heightened levels of awareness the norm. The BCT implements survivability, operations security (OPSEC), and antiterrorism tasks at all fixed locations to maximize protection. The close proximity of civilians and Soldiers can also promote health issues (such as communicable disease) through close contact with local civilians, detainees, or local foods. The protection of civil institutions, processes, and systems required to reach the end state conditions of the stability operations strategy can often be the most decisive factor in stability operations because its accomplishment is essential for long-term success. Civil areas typically contain structured and prepared routes, roadways, and avenues that can canalize traffic. This can lead to predictable friendly movement patterns that can easily be templated by the enemy. An additional planning consideration during stability operations is to protect the force while using the minimum force consistent with the approved rules of engagement (ROE). The escalation of force TTP must also be rehearsed and flexible enough to change with the local threat conditions. Information engagement is a key protection enabler during stability operations. Leaders and Soldiers engage the public to deliver friendly messages and themes (matched by actions on the ground) to key leaders and population groups. Additional protection considerations during stability operations include:

- Reducing the UXO and mine threat in the AO.
- Fratricide prevention and minimizing escalation of force incidents through combat, civilian, and coalition identification measures.
- Developing rapid and efficient personnel recovery techniques and drills.
- Clear OPSEC procedures that account for the close proximity of civilians, NGO, and contractors.
- Disciplined information protection techniques to preserve access to computer networks.
- Containment of toxic chemicals and materials present in the civilian environment.
- Survivability requirements for static facilities, positions, or outposts.

4-53. The presence of any environmental contamination must be identified. The contamination may have existed prior to U.S. forces operating in the area, yet may pose a direct threat to the population that they are not prepared to handle. In this case, it may fall on the engineers, who are the Army's proponent for the environmental mission, to step in and assist as part of the tasks related to stability operations. An example would be environmental contamination that threatens the water supply. It may also be that the environmental contamination identified is directly related to or caused by the actions of U.S. forces. This poses a liability for the U.S. and would need to be mitigated. Prior to U.S. forces occupying a base camp, or within 30 days of occupation, an environmental baseline survey is to be conducted to identify the existence and extent of environmental contamination in the area. This is to protect the health and well-being of Soldiers and to help to limit liability issues for the U.S.

## COMMAND AND CONTROL

4-54. Stability operations are difficult and may require unique skills by staff officers and noncommissioned officers (NCO). Staff members may need to be reorganized to perform the specialized responsibilities for information engagement and contracting tasks. Rank might not indicate the required talent. In stability operations, a few Soldiers under a talented junior NCO doing the right things can succeed, while a larger force doing the wrong things may fail.

4-55. A force optimized for stability operations should have cultural advisors. The current force structure gives corps and division commanders a political or cultural advisor. The BCT commander should consider selecting a political and cultural advisor as a major additional duty. This person may or may not be a commissioned officer. The position requires someone with good interpersonal skills and an understanding of the local environment. Commanders should not try to be their own cultural advisor. They must be fully aware of the cultural dimension, but this is a different role. In addition, this position is not suitable for intelligence professionals. They assist, but their task is to understand the environment. The cultural advisor's job is to help understand and then shape the local conditions with key influencers in the area of operations.

4-56. The omnipresence and global reach of today's news media influences the conduct of military operations more than ever before. Satellite receivers are common even in developing countries. Bloggers and print, radio, and television reporters monitor and comment on everything military forces do. Insurgents use terrorist tactics to produce images that they hope will influence public opinion both locally and globally. The BCT should develop public affairs battle drills to respond to significant activity in its AO and defeat an enemy's advantage by getting information out to the local populace and world stage. BCT commanders must routinely engage with local, friendly, adversary, and international media representatives.

4-57. Leaders must train Soldiers to consider how the global audience might perceive their actions. Soldiers must assume that the media will publicize everything that they say or do. Treat the media as an ally and resource. Assist reporters to meet their information requirements. This helps them portray military actions favorably and reduces misinformation and information fratricide. Good relationships with non-embedded media, especially host nation media, can dramatically increase the BCT's situational awareness as well.

4-58. The requirement to visit the sites of many ongoing stability tasks in an unsecure environment makes aviation a key capability for battle command in stability operations. When aircraft are available, the BCT commander prioritizes their use to accomplish this. The S-3, S-6, and brigade aviation element (BAE) coordinate aerial command post capabilities.

## ASSESSMENT OF STABILITY OPERATIONS

4-59. Continuous assessment is essential to stability operations. Leaders must develop measures of effectiveness, change indicators during mission analysis, and continuously refine them as the operation progresses. Leaders should base these measures and their indicators on mission analysis and course of action (COA) criteria used during the military decision-making process (MDMP). Leaders can also use the operational variables: political, military, economic, social, information, infrastructure plus physical environment and time (PMESII-PT) (FM 3-0). Leaders should use them to develop an in-depth operational picture. They must understand how the operation is changing, not just that it is starting or ending.

4-60. Typical measures of effectiveness and change indicators include the following:

- Percentage of engagements initiated by friendly forces versus those initiated by insurgents.
- Longevity of friendly local leaders in positions of authority.
- Number and quality of tips on insurgent activity that originate spontaneously.
- Economic activity at markets and shops.
- Amount of tips or actionable information received from the civilian populace.
- Number of requests filled for construction projects, services, or supplies.
- Changes in indirect fire support.
- Mean time between civic services failures (e.g., power outages).
- Local crime statistics.

## SECTION IV – INTERAGENCY, INTERGOVERNMENTAL, AND NONGOVERNMENTAL ORGANIZATIONS

4-61. A BCT participating in stability operations works with a wide variety of organizations, especially as the stability operations framework shifts from violent conflict toward normalization. Higher headquarters' directives, Army regulations, U.S. law, host nation law, international treaties, and The Hague and Geneva Conventions dictate and regulate the interaction between the BCT and various organizations. Participation of these organizations requires forging a comprehensive approach, with a shared understanding and appreciation for the intended end state. This approach is both the overall goal and the greatest challenge to mission accomplishment. Many organizations cannot be compelled to work with the BCT, nor do they have any incentive to do so. Therefore, the BCT must build strong relationships through cooperation and coordination. Table 4-1 summarizes some of the organizations the BCT may work within its AO. FM 3-07 provides additional information on interagency, intergovernmental, and nongovernmental organizations.

**Table 4-1. Organizations a BCT may work within an AO**

| |
|---|
| *Interagency Organizations*<br>Interagency organizations are U.S. government agencies and departments that work together to achieve an objective (Joint Publication [JP] 3-0). Examples of such organizations are:<br>Department of State (DOS).<br>Department of Justice.<br>U.S. Agency for International Development.<br>Central Intelligence Agency.<br>Federal Bureau of Investigation.<br>Provincial reconstruction teams. |
| *Intergovernmental Organizations*<br>Intergovernmental organizations are created by a formal agreement (e.g., a treaty) between two or more governments. The organization may be established on a global, regional, or functional basis for wide-ranging or narrowly defined purposes. The purpose of these organizations is to protect and promote national interests shared by member states (JP 3-08). Examples include:<br>United Nations.<br>European Union.<br>Treaty organizations such as the North Atlantic Treaty Organization (NATO). |
| *Nongovernmental Organizations*<br>Nongovernmental organizations are private, self-governing, not-for-profit organizations. They are dedicated to an or all of the following: alleviating human suffering; and/or promoting education, health care, economic development, environmental protection, human rights, and confliction resolution; and/or encouraging the establishment of democratic institutions and civil society (also see JP 3-08). Examples include:<br>The International Red Cross and Red Crescent Movement.<br>Relief organizations, such as Oxfam and World Vision. |
| *Contractors*<br>Contractors to various U.S. or other JIIM organizations support a wide array of stability tasks. Local commanders, other Army headquarters, or interagency and intergovernmental organizations may contract them to work on projects in the BCT AO. Civilian contractor organizations may require security or sustainment. The rules for interaction, support, and authority must be clearly delineated with contractor organization leaders and incorporated into BCT operating procedures and orders. When working closely with the BCT, they may participate in a unit's CMOC to enable better coordination. Examples of contractors support include:<br>Maintaining facilities and services for military or host nation organizations.<br>Conducting civil support work, such as power generation, trash collection, providing water.<br>Providing transportation services.<br>Providing local security services.<br>Maintaining technical equipment for U.S. or host nation forces. |

## SECTION V – TRANSITIONS

4-62. Stability operations include transitions of authority and control among military forces, civilian agencies and organizations, and the host nation. Each transition involves inherent risk. That risk is amplified when the force must manage multiple transitions simultaneously, or when the force must conduct a series of transitions quickly. Planning anticipates these transitions, and careful preparation and diligent execution ensures they occur without incident. Transitions are identified as decisive points on lines of effort. They typically mark a significant shift in effort and signify the gradual return to civilian oversight and control of the host nation.

4-63. An unexpected change in conditions may require the BCT commander to direct an abrupt transition between phases. In such cases, the overall composition of the force remains unchanged despite sudden changes in mission, task organization, and rules of engagement. Typically, the BCT task organization evolves to meet changing conditions; however, transition planning must also account for changes in the mission. Commanders attuned to sudden changes can better adapt their forces to dynamic conditions. They continuously assess the situation and task-organize, and they cycle their forces to retain the initiative. They strive to achieve changes in emphasis without incurring an operational pause.

## PREPARING FOR HANDOVER

4-64. Planning for handover starts when the BCT assumes the stability mission. When a BCT conducting stability operations is relieved, it is likely that the requirement for continuing stability operations will continue. There will be a relief in place and transfer of authority, and the relieving unit will need as much knowledge as the BCT can provide.

4-65. Folders, files, paper and/or digital information should be available for handover. Such information includes lessons learned, details about the populace, village and patrol reports, updated maps, and photographs—anything that will help newcomers master the environment. Computerized databases are fine. If these are not available from the unit being relieved, the BCT should start handover folders in every platoon and specialist squad immediately upon arrival. In addition, it should keep good back-ups and ensure that a hard copy of key artifacts and documents exists. Handover folders reduce the loss of momentum that occurs during any handover by ensuring that information on local conditions is relayed at the lowest levels of command.

## ENDING THE MISSION

4-66. As the end of the stability mission for the BCT approaches, the key leadership challenge becomes keeping the Soldiers focused. They must not reduce their security. They must continue to monitor and execute the many programs, projects, and operations. It is important to safeguard information concerning transition activity. The local people know that Soldiers are leaving and probably have a good idea of the generic transition plan. They have seen units come and go. However, details of the transition plan must be protected; otherwise, the enemy might use the handover to undermine any progress made during the tour. The BCT must help to ensure the follow-on unit's success by maintaining operational security during the handover.

## SECTION VI – SECURITY FORCE ASSISTANCE

4-67. Army doctrine defines "security force assistance" (SFA) as the unified action taken to generate, employ, and sustain local, host-nation, or regional security forces in support of a legitimate authority (FM 3-07). SFA is part of the FM 3-0 construct of full spectrum operations. BCTs conduct SFA across the spectrum of conflict within any of the operational themes. Usually, SFA is part of a larger security sector reform effort. SFA activity can support BCT civil security and civil control stability tasks. For more detail on SFA for the BCT, see FM 3-07.1.

4-68. Because the BCT can operate in nonpermissive and permissive environments, it can conduct SFA across the spectrum of conflict. It supports civilian, military, joint and multinational actors. Examples of the range of possible support include movement security, sustainment, augmentation of reconstruction teams and/or security for elections. BCTs conducting SFA may support foreign security force (FSF) development, assist FSF operations, and support and assist the development of host nation institutions and infrastructure. While providing this support to the host nation, the BCT remains capable of conducting full spectrum operations simultaneously.

## SECURITY FORCE ASSISTANCE FRAMEWORK

4-69. FM 3-07.1 describes the SFA framework as consisting of the mindset required of BCT units and Soldiers, imperatives for success, inherent tasks and activities, and the three types of SFA. The following paragraphs provide a short summary of each element in the framework.

## IMPERATIVES

4-70. The SFA imperatives are:

- Understand the operational environment.
- Provide effective leadership.
- Building legitimacy.
- Manage information.
- Ensure unity of effort.
- Sustain the effort.

4-71. These imperatives come from the historical record and recent experience and provide a focus for the BCT to successfully conduct security force assistance.

## TASKS

4-72. There are five SFA tasks:

- **Organize.** All measures taken to assist foreign security forces to improve their organizational structure, processes, institutions, and infrastructure.
- **Train.** Assistance to foreign security forces to develop programs and institutions to train and educate their forces based on their security environment. This also includes training the U.S. trainers.
- **Equip.** All efforts to assess and assist the foreign security forces with the procurement, fielding, and sustainment of equipment.
- **Rebuild and build.** Efforts undertaken to assess, rebuild, and build the existing capabilities and capacities of foreign security forces and their supporting infrastructure.
- **Advise and assist.** The BCT works with foreign security forces to improve their capability and capacity.

## TYPES OF SECURITY FORCE ASSISTANCE

4-73. BCT commanders use the three types of SFA—advising, partnering, or augmenting—to accomplish the mission. The exact nature of a unit's assigned mission, the operational variables, and the mission variables drive modifications to headquarters and maneuver units. For example, a commander may need to task one unit with advising, task another unit with partnering, and yet another unit with augmenting. In this case, the unit tasked to advise would provide teams for the appropriate FSF. The unit that is partnering could provide staff officers to assist the FSF headquarters. The third unit could provide squads to augment FSF platoons. Tasked units request specialized support, sustainment, and medical support as required.

# AUGMENTATION

4-74. The BCT receives augmentation, based on the requirements of the operational environment, with enabling assets and capabilities to support distributed SFA. Any of the three BCTs, heavy, Infantry, or Stryker, can support SFA. Based on the overall mission analysis, including an analysis of functions and requirements, commanders and staffs determine how to task organize. No standard task organization exists since conditions vary so widely. Because there is not standard task organization, there is also no corresponding standard equipment list. Units must be prepared to generate mission essential equipment lists and/or operational needs statements in support of the SFA mission. Force tailoring includes providing BCTs with additional forces, personnel, or capabilities (e.g., the possibility of an embedded provincial reconstruction team). Additional assets and capabilities can include command and control, communications, sustainment, engineer, military police, and intelligence.

4-75. The military transition team is a key subordinate unit for the brigade conducting security force assistance. The military transition team mission is to assist foreign security force military units. When the brigade provides organic forces to form the basis for the brigade military transition team, the company team is the foundation of a brigade military transition team. Additional personnel and assets augment the company team. The BCT must facilitate the operations of military transition teams, such as protection, transportation, sustainment, and communications.

# Chapter 5

# Security Operations

Security is an essential part of full spectrum operations. Although not an element of full spectrum operations, security operations are inherent in offensive, defensive, stability, and civil support operations. The ultimate goal of security operations is to protect, prevent surprise, and reduce the unknowns in any situation (FM 3-90).

This chapter discusses the Brigade Combat Team's (BCT) performance of security operations. This chapter details the five forms of security operations. They are screen, guard, cover, area security and local security.

## SECTION I – OVERVIEW

## PURPOSE

5-1. The purpose of security operations is to provide information that gives the main body commander the reaction time and maneuver space needed to fight the enemy effectively. Generally, the BCT assigns security missions to its reconnaissance squadron, but brigade-size security operations may require that all subordinate elements participate. Units assigned security missions must provide information about the enemy and terrain, prevent the main body from being surprised, and preserve the combat power of friendly forces for decisive employment. Critical information includes the enemy's size, composition, location, direction, and rate of movement. Terrain information focuses on obstacles, avenues of approach, and key terrain features that impact the movement of either force. The BCT and higher commander may assign other CCIR.

5-2. The BCT performs security operations as part of full spectrum operations. While conducting offensive or defensive actions, security operations are designed to provide early warning, protect the force and enable the BCT commander to retain the initiative and freedom of maneuver. In stability and civil support operations, security operations focus on protecting civilians, protecting the force, securing services, and safeguarding relief/recovery operations and/or national building efforts. The nature of the security mission, the organic composition of the security force, and the enemy situation determine what augmentation the BCT needs. The BCT assigns security tasks or missions to its reconnaissance squadron; sustained security tasks or missions usually require participation by the entire BCT. When the BCT assigns a security task or mission to a subordinate element, the BCT ensures the subordinate element is task organized and has been allocated the resources to meet mission requirements. Types of forces allocated could include tank and mechanized Infantry units, reconnaissance units, engineer elements, attack helicopter units, close air support (CAS) priority, and intelligence systems.

## RECONNAISSANCE

5-3. Reconnaissance is inherent and continuous in all security operations. The focus of reconnaissance is preventing the surprise of the protected force commander. Reconnaissance provides information that allows the commander to make decisions regarding maneuver and fires, and provides reaction time to implement those decisions. Unmanned aircraft systems (UAS) and ground scouts and sensors are synchronized to maximize their complementary capabilities.

## COUNTERRECONNAISSANCE

5-4.  Counterreconnaissance is also inherent in all security operations. Counterreconnaissance is the sum of all actions taken at each echelon to counter enemy reconnaissance and surveillance efforts throughout the area of operations (AO). Its purpose is to deny the enemy information about friendly operations and/or to destroy or repel enemy reconnaissance elements. security forces operate either offensively or defensively when executing counterreconnaissance. The BCT's designated counterreconnaissance plan provides the active and passive measures to defeat the enemy's reconnaissance efforts and protect the friendly force from observation.

## SECTION II – FUNDAMENTALS AND PLANNING CONSIDERATIONS

## FUNDAMENTALS OF SECURITY

5-5.  Per FM 3-90, the five fundamentals for planning and performing successful security operations are:
- Orient on the force, area, or facility to be protected.
- Perform continuous reconnaissance.
- Provide early and accurate warning.
- Provide reaction time and maneuver space.
- Maintain enemy contact.

## PLANNING CONSIDERATIONS

5-6.  Security operations are conducted to collect, analyze, and provide intelligence information to the supported commander, enabling him time to plan, prepare, and/or deploy against expected or unexpected enemy activities. These operations vary by the type of combat and types of terrain (open/rolling to complex/urban) the BCT encounters.

### ENGAGEMENT/DISPLACEMENT CRITERIA

5-7.  Engagement criteria specify those circumstances (by unit/element) for initiating engagement with an enemy force. Conversely, the security force commander's understanding of what the BCT commander requires (or expects the security force to destroy), and his understanding of the enemy's most likely course of action, enables him to identify the unit's engagement criteria. This enables unit leaders to focus certain weapons systems or to develop engagement areas and plan for the destruction of specified enemy elements if encountered. Displacement criteria relates to engagement criteria. The BCT commander defines what events, or triggers, cause the security force to reposition or hand off responsibilities. Examples of such causes are a certain size force, or a specific enemy formation, or an element reaching a given point or graphic control measure. security force commanders should use surveillance assets to assist in maintaining contact and/or executing hand over to follow-on forces.

5-8.  The security force commander determines:
- Specified and implied tasks based upon higher commander's guidance.
- Critical security tasks to be performed by subordinate units.
- Task organization for security, command and support relationships, and command and control (C2) structure.
- Actions on contact.
- Potential branches and sequels to the operation.
- Communications plan (architecture and required support).
- Available intelligence and collection assets at the joint and national levels.

## SECTION III – SCREEN

5-9. The primary purpose of a screen is to provide early warning to the main body. Based on the higher commander's intent and the unit's capabilities, it might also destroy enemy reconnaissance, and impede and harass the enemy main body with indirect and/or direct fires. Screen missions are defensive in nature and largely accomplished by establishing a series of observation posts (OP) and conducting patrols to ensure adequate surveillance of the assigned sector. The screen provides the protected force with the least protection of any security mission. This mission is appropriate when operations have created extended flanks, when gaps exist between major subordinate maneuver units that cannot be secured in force, or when required to provide early warning over gaps that are not considered critical enough to require security in greater strength. This permits the main body commander to maximize the security effort where contact is expected.

5-10. The BCT usually conducts screen missions with just its organic assets. However, it may be augmented with additional assets depending on the division commander's intent, or on the mission enemy terrain and weather, troops and support available, time available, and civil considerations (METT-TC). Usually, the trace of the screen is established within the range of main body artillery. However, situations can require operating beyond that range and may require locating artillery in direct support of the screening unit.

5-11. Because a screen is defensive in nature, a screen may be performed for a stationary force to the front, flank, or rear of the main body. A screen is performed for a moving force to the flank or rear of the main body. A screen mission is not performed forward of a moving force. Zone reconnaissance is more suited for operations forward of a moving force.

5-12. Displacement of the screen to subsequent OP positions is event-driven. The approach or detection by an enemy force, relief by a friendly unit, or movement of the protected force dictates screen movement. Displacing the screen, executed by well-rehearsed security drills performed at the platoon and company/troop levels, provides security and maintains contact for the security force as it displaces. The main body commander does not place a time requirement on the duration of the screen unless the intent is to provide a higher level of security to the main body or to provide a tentative time frame for subordinate unit planning purposes.

# CRITICAL TASKS

5-13. A screen mission has certain critical tasks that guide planning. The level to which the unit can achieve these critical tasks is dependent on the unit's capabilities, the commander's intent, and METT-TC. To achieve the intent of a screen mission, units must complete the following critical tasks:

- Maintain continuous surveillance of all avenues of approach that affect the main body's mission under all conditions. METT-TC may necessitate continuous monitoring of smaller avenues of approach.
- Destroy or repel all reconnaissance elements within capabilities (counterreconnaissance).
- Locate the lead elements that indicate the enemy's main attack.
- Maintain contact with the enemy's lead element while displacing and reporting its activities.

# STATIONARY SCREEN

## MAIN BODY COMMANDER'S GUIDANCE

5-14. The main body commander should provide the screening force commander with guidance concerning:

- Augmentation.
- General trace of the screen and time at which the screen must be established.
- Screened frontage.
- Force to be screened.

- Rear boundary of the screening unit.
- Possible follow-on missions.

## Augmentation

5-15. Augmentation is any additional assets that the screening unit receives to conduct the mission. This may include ground maneuver forces, aviation assets, artillery, air defense, engineers, or additional logistical support. Augmentation can include an antiarmor platoon or company, a tank platoon, a reconnaissance platoon, a sniper squad, an Infantry company, an engineer platoon, or additional UAS and unmanned ground sensors (UGS).

## Screen Trace and Timeline

5-16. A phase line (PL) placed along identifiable terrain graphically indicates the trace. Units should take into consideration the amount of early warning, range of indirect fires, desired main body maneuver space, and fields of observation. When screening forward of the BCT, this PL represents the forward line of own troops (FLOT), and could be located along or close to a coordinated fire line. Placing screening units beyond the trace line requires approval of higher headquarters, and usually requires modification of fire support control measures.

## Screened Frontage

5-17. The tasks required of a screening unit are minimal compared to other security missions. Therefore, the screening unit may be assigned a wide frontage. The commander directs or requests augmentation if the subordinate screening force is required to screen beyond its capabilities. Careful consideration must be given when assigning UAS or ground-based sensors, as weather, station time, and terrain can affect the augmentation's ability to execute the mission. Units should use UAS and ground-based sensors to complement ground forces and to provide extended depth, some width, and increased flexibility to the operation.

## Force to be Screened

5-18. The security force must understand the mission, purpose, and commander's intent of the unit it is screening. Knowing this information enables the screen force commander to better focus his elements and enhances initiative during execution.

## Rear Boundary of the Screening Unit

5-19. The rear limit of the screening unit is depicted as a boundary. Responsibility for the area between the screened force and the screening unit rear boundary lies with the screened (main body) force. This boundary reflects time and space requirements, clearly delineates terrain responsibilities, and provides depth required by the screening unit. The boundary may also serve as a battle handover or reconnaissance handover line to control passing responsibility for the enemy to the protected force.

## Possible Follow-on Missions

5-20. To facilitate planning and future operations, the next likely mission the screening unit is to perform should be defined with enough information to enable the commander to begin planning and preparing for it. Providing this information also helps define the end state of the screen mission.

## SCREENING UNIT COMMANDER'S CONSIDERATIONS

5-21. Given the higher commander's guidance, the BCT commander considers several issues:
- The initial screen.
- Movement to occupy the screen.
- Control of displacement to subsequent screen lines.
- AOs for subordinate units.

- Air and ground integration.
- Surveillance and acquisition assets.
- Fire planning.
- Mobility, countermobility, and survivability.
- Command and control.
- Sustainment.

## Initial Screen

5-22. The controlling headquarters establishes the initial screen. It is adjusted closer only with approval from higher headquarters. Because the initial screen often represents the FLOT, it is considered a restrictive control measure. Coordination is required to move beyond the initial screen to conduct aerial surveillance or ground reconnaissance. If operations forward of the screen are required, an additional phase line should be established to designate the screening unit's limit of advance. Key considerations in locating the screen are:

- Range or responsiveness of indirect fire support.
- Limits of detection from behind the screen line.
- Requirements to observe specific named areas of interest (NAI) or targeted areas of interest (TAI).

## Movement to Occupy the Screen

5-23. Time and the enemy situation determine the method of occupying the screen. There are three primary methods available to occupy the screen:

- Zone reconnaissance.
- Infiltration.
- Tactical road march.

5-24. If the situation is vague, or if more information is required on the terrain between the main body and the screen, and if time is available, the BCT can conduct zone reconnaissance to the screen lines. This method identifies any enemy in the AO and familiarizes the unit with the terrain. It is time-consuming but provides the most security.

5-25. If the enemy situation is vague, or if the enemy is known to be in the AO, and if the intent is not to make contact with the enemy prior to occupying the screen, the screening unit should conduct infiltration to get to the screen. Infiltration provides the optimum level of stealth; however, it is time-consuming and less secure for the unit due to the reduction of flexibility in massing combat power. If there is an accurate picture of the enemy situation or if time is short, the screening units may conduct tactical road marches to positions just short of their screen lines.

## Control of Displacement to Subsequent Screen Lines

5-26. The screening unit's commander uses phase lines to control the operation. Since displacement to subsequent positions is event-driven, subsequent phase lines serve to guide the unit commander's initiative during the mission. The plan should define the event criteria triggering displacement, and displacement should be controlled at unit levels.

## Area of Operations for Subordinate Units

5-27. The commander designates AOs for the subordinate units. Terrain responsibility for NAIs and TAIs goes with the AO. Units usually are deployed with UAS and/or UGS, and Prophet systems positioned to provide depth for the screening unit. Reduced depth is the trade-off when screening extends frontages. When forced to do so, the commander may have to task UAS or sensors to monitor terrain. This terrain should not be high-speed avenues of approach. Plans must include redundancy of coverage and compensate for the absence of UAS (e.g., in the case of adverse weather) by adjusting subordinate AOs or augmentation.

### Air and Ground Integration

5-28. UASs can conduct surveillance forward, to the rear, or to the flanks of the screening units to add depth and extend the capabilities of the ground screen. They can conduct surveillance along an exposed flank of the screening unit; assist in patrolling gaps between units; augment surveillance of NAIs; and add depth within the AO along subsequent screens. The concepts of battle or reconnaissance handover are used within the unit as aerial and screening elements displace to subsequent lines or positions. This ensures that the unit maintains contact with the enemy. The BCT or subordinate battalions can control external UAS augmentation In either case, integrating air and ground assets greatly enhances the effectiveness of the screen.

### Surveillance Assets

5-29. The BCT should plan the integration of surveillance and target assets to provide the earliest possible warning on the most likely enemy activity using its organic assets and/or external assets. The BCT can request supporting Army and joint interagency, intergovernmental, and multinational (JIIM) assets through its higher headquarters. These assets can provide initial acquisitions (e.g., Joint Surveillance Target Attack Radar System [JSTARS] or Guardrail moving target indicator acquisition) that cue organic assets such as UAS and ground observers to obtain more definitive intelligence and continue tracking the enemy. Units also can plan these assets to cover the screening unit while it collapses the screen and is most vulnerable, or to assist in regaining contact with the enemy if contact is lost. If the screening unit is screening extended frontages, these assets can operate in an economy of force role by conducting periodic surveillance of areas the enemy is less likely to use but still has the possibility of using. The sensors organic to the BCT can be used under BCT control or tasked to subordinate units. In either case, integrating the sensors into the plan greatly enhances the effectiveness of the screen.

### Indirect Fire Planning

5-30. Indirect fire planning integrates artillery, rockets, mortars, aviation and close air support. Planning starts with the BCT's fire support plan, which includes assigned fires, support tasks. The planning must consider the higher and BCT commanders' intent for the screen. For example, the commander's intent may be to report and maintain contact only; or it may be to delay or to destroy specific elements of the enemy's formations. Targets should be planned on likely approaches at choke points or areas where the enemy must slow down. The BCT should plan indirect and overwatching direct fires in conjunction with any obstacles the screening unit emplaces. It should also plan engagement areas to focus fires at points along likely enemy avenues of approach. This is where it is most possible to achieve the desired effects. It is critical that the BCT clearly identifies what supporting artillery and munitions are available to the screening unit. Also critical for the BCT to identify clearly are the command relationship, their tactical mission, the communications/digital linkages, and artillery positioning plans.

### Assured Mobility and Survivability

5-31. Engineers may augment subordinate screening units for specific tasks. Typical engineer tasks are survivability and assured mobility of the screening force, which includes the emplacement of situational obstacles. Situational obstacles are obstacles that the BCT plans, and possibly prepares, but does not execute until specific criteria are met. Therefore, units may or may not execute situational obstacles depending on the situation that develops during the battle. They are "be prepared" obstacles and provide the commander with flexibility for emplacing tactical obstacles based on battlefield development. In screen operations, situational obstacles can be used to disrupt and delay the enemy (in conjunction with indirect and direct fires), and to protect elements of the BCT.

### Command and Control, and Sustainment

5-32. In most instances, both the tactical command post and main command post (CP) must be operational to support C2 over extended distances and to maintain digital linkages with BCT headquarters (HQ) and its subordinate elements. Initial and subsequent locations of the main CP must be integrated into the BCT's communications plan to ensure that continuous digital connectivity is maintained. Sustainment assets should be prepared for operations extended in both time and space. Screening assets operating well forward of, or to the flank of, the BCT may need support from the closest maneuver units. The need for this support must be determined early in the planning process to allow the brigade sustainment battalion time to plan and position assets to provide support to the extended BCT screening assets.

# MOVING SCREEN

5-33. The same planning considerations discussed above apply to a moving screen. Emphasis may shift since the main body is moving. The BCT might require the reconnaissance squadron or maneuver unit to conduct moving flank screens and potentially screen the rear of the main body as it attacks. Screening the rear of a moving force is essentially the same as a stationary screen. As the protected force moves, the screening unit occupies a series of successive screens. Movement is regulated by the requirement to maintain the time and distance factors desired by the main body commander. UAS or sensors may assume the screen during movement of ground troops or work to extend the areas of coverage.

5-34. The moving flank screen poses additional considerations. The width of the screen frontage is not as important as the force being protected and the enemy avenues of approach that might affect the main body's movement. The unit screens from the front of the lead combat element of the main body to the rear of the protected elements (excluding front and rear security forces). The combat trains moves with the screening unit, and the field trains moves with the brigade support battalion (BSB).

5-35. Command and control is challenging during a moving screen because the unit moves in one direction and orients in another. Control measures must facilitate both orientations. For example, phase lines serve as on-order unit boundaries and do not divide avenues of approach into the flank of the main body. The unit plans not only for the advance and initial screen, but also for a screen in depth back to the main body.

5-36. The speed of the main body, distance to the objective, and the enemy situation determine movement along the screen. Unit movement centers on a designated route of advance. This route is parallel to the axis of advance of the protected force, and large enough to accommodate rapid movement of the unit and facilitate occupation of the screen. The route must be kept clear to ensure rapid movement of the augmenting, sustainment, and C2 assets. These elements should stay off the main route unless moving, or travel on alternate routes.

## SECTION IV – GUARD

5-37. A commander employs a guard when he expects enemy contact and requires additional security beyond that provided by a screen. The multiple requirements of the guard mission often are performed simultaneously over relatively large areas. The unit requests guidance on the priority of tasks. If the unit determines that it cannot complete its assigned task after starting the guard, it must report this to the commander and await further instructions.

5-38. The three types of guard operations are advance, flank, and rear guard. A commander can assign a guard mission to protect either a stationary or a moving force. Guard tasks include:

- Maintain contact with its main body and any other security forces operating on its flanks.
- Destroy or fix the enemy force.
- Maintain contact with enemy forces and report activity in the AO.
- Maintain continuous surveillance of avenues of approach to the AO under all visibility conditions.
- Impede and harass the enemy within its capabilities while displacing.
- Cause the enemy to deploy and then report its direction of travel.
- Permit no enemy ground element to pass through the security area undetected and unreported.
- Destroy or cause the withdrawal of all enemy reconnaissance patrols.

5-39. A guard force contains sufficient combat power to defeat, cause the withdrawal of, or fix the lead elements of an enemy ground force before the enemy force can engage the main body with direct fire. This is one of several ways in which a guard differs from a screen. A guard force routinely engages enemy forces with direct and indirect fires. A screening unit, on the other hand, primarily uses indirect fires or CAS to destroy enemy reconnaissance elements and slow the movement of other enemy forces. A guard force uses all means at its disposal to prevent the enemy from penetrating to a position where it could observe and engage the main body. A guard force operates within the range of the main body's indirect fire systems and deploys over a narrower front than a screening force of comparable size to concentrate combat power.

# ADVANCE GUARD

5-40. An advance guard for a stationary force is defensive in nature. It defends or delays in accordance with the main body commander's intent. An advance guard for a moving force is offensive in nature. The advance guard develops the situation so the main body can use its combat power to the greatest effect. The main body's combat power must not be consumed, reinforcing the advance guard. The full combat power of the main body must be available immediately to defeat the main enemy force.

5-41. A BCT advance guard for a moving force usually conducts a movement to contact (Figure 5-1). It task organizes and uses the graphics of a movement to contact. The advance guard conducts shaping operations and is not the main effort or decisive operation for the higher headquarters. Ground subordinate elements of an advance guard usually deploy abreast to cover the axis of advance or the main body's AO. The advance guard is responsible for clearing the axis of advance or the designated portions of the enemy elements' AO. This enables the main body to move unimpeded, prevents unnecessary delay of the main body, and defers deployment of the main body for as long as possible.

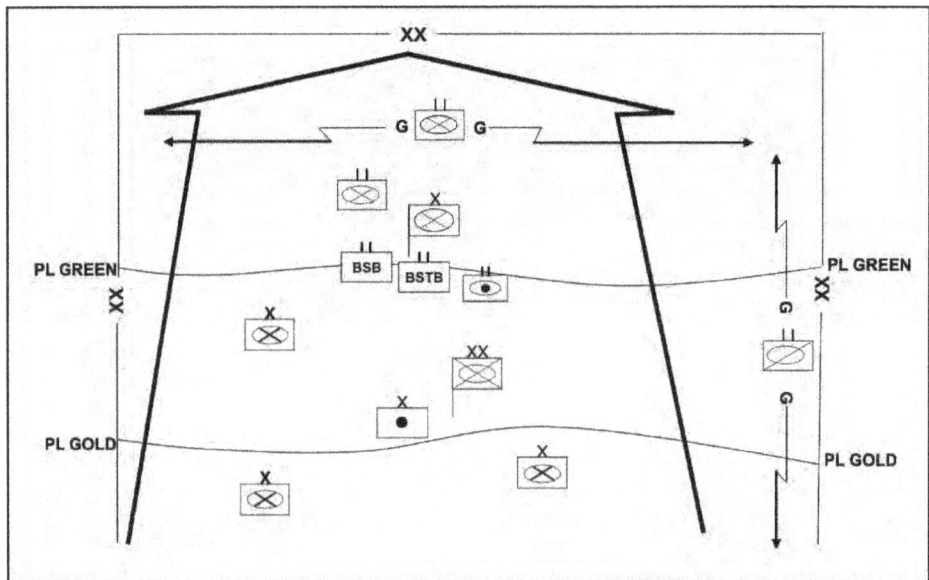

Figure 5-1. HBCT as a division advance guard

# FLANK GUARD

5-42. A flank guard protects an exposed flank of the main body. A flank guard is similar to a flank screen except the commander plans defensive positions in addition to screen lines. The flank guard is responsible for clearing the area from the supported main body to the flank guard's designated positions. Usually this area extends from the forward screen, along the flank of the formation, to either the front line trace of troops or the rear of the moving formation, tying in with the rear guard.

# REAR GUARD

5-43. The rear guard protects the exposed rear of the main body. This occurs during offensive operations when the main body breaks contact with flanking forces or during a retrograde. The commander may deploy a rear guard behind both moving and stationary main bodies. The rear guard for a moving force displaces to successive battle positions along PLs or delay lines in depth as the main body moves. The nature of enemy contact determines the exact movement method or combination of methods used in the displacement (successive bounds, alternate bounds, and continuous marching).

## SECTION V – COVER

# COVERING FORCE

5-44. A covering force is a self-contained force capable of operating independently of the main body, unlike a screening or guard force. A covering force, or portions of it, often becomes decisively engaged with enemy forces. Therefore, the covering force must have sufficient combat power to engage the enemy and accomplish its mission. A covering force develops the situation earlier than a screen or a guard force. It fights longer and more often and defeats larger enemy forces.

5-45. The covering force's distance forward of the main body depends on the intentions and instructions of the main body commander; the terrain; the location and strength of the enemy; and the rates of march of both the main body and the covering force. The width of the covering force area is the same as the AO of the main body.

5-46. While a covering force provides more security than a screen or guard force, it also requires more resources. Before assigning a cover mission, the higher commander must ensure that the BCT has sufficient combat power to resource a covering force, while maintaining enough for the decisive operation. When the commander lacks the resources to support both, he must assign his security force a less resource intensive security mission; that is, either a screen or a guard.

## DEFENSIVE COVERING FORCE

5-47. A defensive covering force prevents the enemy from attacking at the time, place, and combat strength of his choosing. Defensive cover gains time for the main body enabling it to deploy, move, or prepare defenses in the main battle area (MBA). It accomplishes this by disrupting the enemy's attack, destroying his initiative, and establishing the conditions for decisive operations. The covering force makes the enemy deploy repeatedly to fight through the covering force and commit his reserve or follow-on forces to sustain momentum (Figure 5-2).

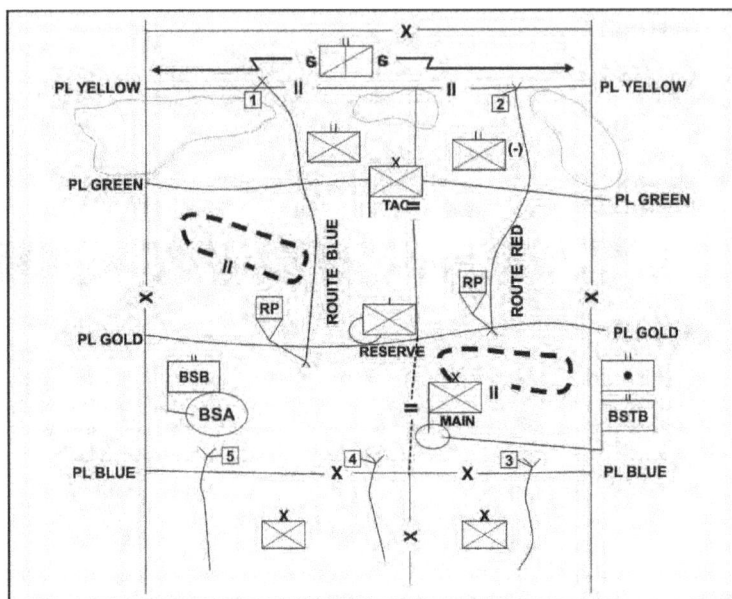

Figure 5-2. Example of covering force plan

MISSION

5-48. A covering force accomplishes all the tasks of screening and guard forces. A covering force for a stationary force performs a defensive mission, while a covering force for a moving force generally conducts offensive actions. A covering force usually operates forward of the main body in the offense or defense, or to the rear for a retrograde operation. Unusual circumstances could dictate a flank covering force, but this is usually a screen or guard mission. An offensive covering force seizes the initiative early for the main body commander enabling him to attack decisively. When the main body commander perceives a significant enemy to one of his flanks, he usually establishes a flank covering force. That force conducts its mission in much the same way as a flank guard performs its mission. The main differences between the two missions are the scope of operations and the distance the covering force operates away from the main body.

5-49. Just as in a flank guard, the flank covering force must clear the area between its route of advance and the main body. It must also maintain contact with an element of the main body specified by the main body commander. This element usually is part of the advance guard for the flank unit of the main body.

## SECTION VI – SECURITY

# AREA SCEURITY

5-50. Area security is a form of security that includes reconnaissance and security of designated personnel, airfields, unit convoys, facilities, main supply routes, lines of communication, equipment, and critical points. Area security operations are conducted to deny the enemy the ability to influence friendly actions in a specific area, or to deny the enemy use of an area for its own purposes. This can entail establishing and occupying a 360-degree perimeter around the area being secured, or taking actions to destroy enemy forces already present in the area. The area to be secured can range from specific points (e.g., bridges, defiles) to areas such as terrain features (e.g., ridgelines, hills) to large population centers and adjacent areas. Area security requires a variety of shaping operations that include reconnaissance, defensive, offensive, stability, and support tasks. Units can conduct area security in support of any military operation.

## MISSION

5-51. An area security force neutralizes or defeats enemy operations in its assigned area. It screens, reconnoiters, attacks, defends, and delays as necessary to accomplish its mission. Area security operations may be offensive or defensive in nature. They focus on the enemy, the force being protected, a protected asset, or any combination of these. Commanders may balance the level of security measures with the type and level of threat posed in the specific area; however, all-around security is essential (Figure 5-3).

5-52. When conducting an area security mission, the area security force conducts reconnaissance and security operations, and attacks, defends, and delays as necessary to accomplish its mission. Security forces prevent enemy ground reconnaissance elements from directly observing friendly activities within the area being secured. They prevent enemy ground maneuver forces from penetrating those defensive perimeters established by the commander. The commander can have his elements employ a variety of techniques such as OPs, battle positions, ambushes, and combat outposts to accomplish this security mission. His reserve enables him to react to unforeseen contingencies. With available intelligence gathering capabilities, the screening unit has the potential to be more offensive oriented conducting ambushes and preemptive strikes with greater precision to maintain security, if such actions are within the BCT commander's intent and appropriate to the overall situation.

## OTHER SUPPORT

5-53. The reconnaissance squadron or maneuver battalion may conduct, or task subordinate units to conduct, the following in support of area security:

- Area, route, or zone reconnaissance.
- Screen.
- Offensive and defensive tasks (within capability).
- Convoy and route security.
- High value asset security (including fixed site security and personal security detachments).
- Combat outposts.
- Patrols with host nation forces.
- The unit conducts stability tasks to support long-term area security, such as:
    - Liaison/negotiation and establishment of civil-military operations centers.
    - Securing activities/projects for civil-military operations.
    - Compliance inspections.
    - Support presence operation (i.e., support company/platoon checkpoints, presence patrols).
- Deliver supplies or render humanitarian aid.
- React to civil disturbance.
- Plan/react to media.
- Leader and Soldier engagements with the local population.

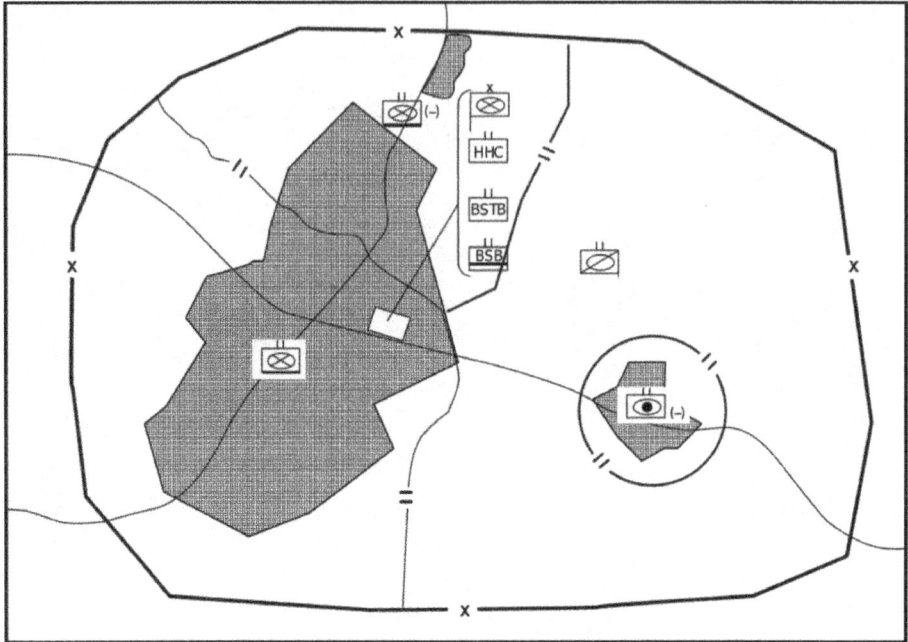

**Figure 5-3. HBCT conducting area security**

5-54. Mission variables determine the augmentation that a BCT might need to execute area security. Particular consideration should be given to the need for the appropriate mix of reconnaissance, maneuver, engineer, and artillery units. External augmentation can include elements of a maneuver enhancement brigade, additional UAS, tactical human intelligence (HUMINT) and counterintelligence teams, civil affairs teams and JIIM, nongovernmental, and contractor elements.

5-55. Focused intelligence preparation of the battlefield (IPB) is vital to supporting area security. The factors of METT-TC and unit capability determine specific unit missions. Some influencing factors are:

- The natural defensive characteristics of the terrain.
- Existing roads and waterways for military lines of communication and civilian commerce.
- The control of land and water areas and avenues of approach surrounding the area to be secured, extending to a range beyond that of enemy artillery, rockets, and mortars.
- The control of airspace.
- The proximity to critical sites such as airfields, power generation plants, and civic buildings.
- Sources external to the BCT AO causing instability in the local population.

5-56. Due to the possibility of commanders tying their forces to fixed installations or sites, these types of security missions may become defensive in nature. This must be carefully balanced with the need for offensive action. Early warning of enemy activity is paramount in area security missions and provides the commander with time to react to any enemy. Maximum use of external collection assets can reduce the requirements placed on BCT reconnaissance assets. Focused reconnaissance planning, dismounted/mounted patrols, and aerial reconnaissance are essential to successful area security.

## OTHER CONSIDERATIONS

5-57. Depth is provided using subsequent fighting positions and mobile reserves. The mobility and firepower of armored forces enable these forces to rapidly traverse large areas and quickly mass and destroy any enemy penetration. The size of the reserve depends on the tactical situation and available forces. Immediate reaction to intelligence information or any type of attack is vital. This immediate reaction to accurate and timely intelligence can enable destruction of enemy elements prior to an attack on the area being secured. Reaction operations or commitments of the reserve are simple, planned, and rehearsed under all the employment conditions possible.

# BASE SECURITY

## OBJECTIVE

5-58. The objective of base security is to maintain a secure position, defending in all directions. The commander can employ base security when conducting the full spectrum of operations. The BCT establishes base security when it must hold critical terrain in areas where the defense is not tied in with adjacent units. The BCT can also form a perimeter and conduct base security when it has been bypassed and isolated by the enemy and must defend in place.

5-59. Forward operating bases can be used to create a 360-degree defense for basing units inside urban environments. These forward operating bases are considered a secure area and most would have guard towers, indirect fire protection, and an infrastructure to support the unit. Many times the BCT houses its support elements in the same forward operating base as the combat unit.

## SUSTAINMENT ELEMENTS

5-60. Sustainment elements can support from inside the perimeter or from another location depending on the mission and status of the BCT. Sustainment considerations are the type of transport available, the weather, and the terrain. Sustainment assets inside the perimeter should be in a protected location from which they can provide continuous support. The availability of drop zones (DZ) and landing zones (LZ) protected from the enemy's observation and fire is a main consideration in selecting and organizing the location.

This page intentionally left blank.

# Chapter 6

# Reconnaissance Operations

"Reconnaissance operations are those operations undertaken to obtain, by visual observation or other detection methods, information about the activities and resources of an enemy or potential enemy, or to secure data concerning the meteorological, hydrographical or geographical characteristics and the indigenous population of a particular area" (FM 3-90). The Brigade Combat Team (BCT) uses its assigned reconnaissance units and surveillance assets to collect information, while using robust capabilities to translate that information into intelligence. This chapter discusses the brigade combat team's execution of reconnaissance operations. This chapter also discusses the processes the staff undertakes to synchronize and integrate information requirements, collection tasks, and available reconnaissance and military intelligence (MI) assets. These factors answer the information requirements that support the commander's understanding and visualization of the operation.

## SECTION I – OVERVIEW

## PURPOSE OF RECONNAISSANCE OPERATIONS IN THE BRIGADE COMBAT TEAM

6-1. Brigade combat teams conduct reconnaissance operations. Through these operations, the BCT obtains the information it needs to develop situational awareness (SA); and it enables the situational understanding (SU) the commander needs to make decisions. The focus of this activity is to answer the commander's critical information requirements (CCIR). Reconnaissance operations are planned and executed early in the BCT's decision-making process (e.g., during mission analysis).

6-2. Intelligence activities enable production of intelligence about the enemy, environment, and civil considerations that the commander needs to make critical decisions. Intelligence products answer CCIR and other information requirements (IR) developed during the operations process. Timely and accurate intelligence encourages initiative, and can facilitate actions that could negate enemy superiority in personnel, materiel or organization. Developing timely and accurate intelligence depends on aggressive and continuous reconnaissance operations.

6-3. The BCT's reconnaissance and surveillance activities are integrated across the intelligence and maneuver warfighting functions. Reconnaissance is a combined arms operation that focuses on priority intelligence requirements while answering the commander's critical information requirements (FM 3-0). The BCT commander and staff continuously plan, task, and employ reconnaissance forces and surveillance systems.

6-4. Reconnaissance and surveillance activities (including MI discipline collection) support the BCT conduct of full spectrum operations through four tasks:
- ISR synchronization.
- ISR integration.
- Surveillance.
- Reconnaissance.

6-5. The BCT's intelligence activities collect, process, store, display, and disseminate information from a multitude of collection sources. Although staff or technical "stovepipes" exist, access to distributed databases is the primary source for obtaining information to produce intelligence. These databases of information exist in higher, lower, and adjacent units. ISR plans are nested from scout platoon through the BCT and higher headquarters. Nesting ensures unity of focus and enables successful execution, layered coverage, and retaining the initiative.

# RECONNAISSANCE SQUADRON

6-6.  The reconnaissance squadron is the main organization that the BCT commander has available for his reconnaissance needs. Reconnaissance squadrons of the Heavy Brigade Combat Team (HBCT), Infantry Brigade Combat Team (IBCT), and Stryker Brigade Combat Team (SBCT) are organized to conduct reconnaissance and security missions throughout the BCT's area of operations (AO). By leveraging information technology and air/ground reconnaissance capabilities in complex terrain, the reconnaissance squadron can develop the situation by focusing on all categories of threats in a designated AO. This enables the BCT commander to maintain battlefield mobility and agility while choosing the time and place to confront the enemy and the preferred method of engagement. The squadron commander has a variety of tools to assist him in conducting reconnaissance and security missions within all spectrums of conflict. He can task organize to optimize complementary effects while maximizing support throughout the BCT's AO (FM 3-20.96).

6-7.  As the "eyes and ears" of the BCT commander, the squadron provides the combat information that enables the BCT commander to develop SU, make better and quicker plans and decisions, and to visualize and direct operations. The squadron progressively builds situational awareness in operational environments that are characterized by combinations of traditional, irregular, disruptive or catastrophic threats or challenges. The squadron employs unique combinations of reconnaissance and security capabilities to successfully meet the information challenges intrinsic to the spectrum of conflict. The squadron's reconnaissance operations yield an extraordinarily high payoff in the areas of threat location, disposition, and composition, early warning, protection, and munitions effectiveness. This preserves the BCT's freedom of maneuver and initiative over the enemy. Skillful reconnaissance operations allow the BCT commander to shape the battlefield, ideally accepting or initiating combat at times and places of his choosing, and applying combat power in a manner most likely to achieve his desired effects.

6-8.  The squadron's primary missions are:
- Reconnaissance.
  - Zone reconnaissance.
  - Area reconnaissance.
  - Route reconnaissance.
- Security.
  - Guard.
  - Screen.
  - Area security.
  - Local security.

## SECTION II – INTELLIGENCE

6-9.  To execute missions effectively, commanders and staff require intelligence about the enemy, terrain, weather, and civil considerations. Intelligence assists commanders in visualizing the operational environment, and assessing operations to achieve the desired end state. Intelligence supports protection by alerting the commander to emerging threats and assisting in security operations. The BCT S-2 has the lead for intelligence planning and synchronizing the execution of intelligence collection within the brigade. The BCT MI company has a critical role in assisting the S-2 in acquiring and analyzing intelligence. Details about the organization of the BCT MI company are provided in Chapter 8. FM 2-19.4 describes the operations of the MI company.

6-10. Details on the intelligence process are found in FM 2-0, FM 2-01.3, and FMI 2-01. Intelligence staff elements in the BCT command posts and MI company, collaborating with lower, higher, and adjacent intelligence staff, produce intelligence products using the intelligence process. These products include:

- Intelligence preparation of the battlefield (IPB) reports and overlays.
- Intelligence running estimate (reports and overlays).
- ISR synchronization matrix and ISR synchronization tools (overlay and collection matrix.
- ISR plan (this is a combined effort by the S-2 and S-3).

# INTELLIGENCE PREPARATION OF THE BATTLEFIELD

6-11. Intelligence preparation of the battlefield (IPB) is a systematic and continuous process for analyzing the threat and environment in a specific geographic area. It is a staff planning activity undertaken organized by the S-2 and supported by the entire staff to define and understand the operational environment and the advantages and disadvantages presented to friendly and threat forces. IPB supports each staff section's running estimates and the military decision-making process (MDMP). The S-2 uses IPB to describe the environment in which the brigade is operating and the effects of the environment on brigade operations. The IPB process supports the S-2 in determining threat capabilities, objectives, and courses of action (COA). The S-2, supported by the Analysis and Integration Platoon, conducts IPB prior to and during the brigade's planning for an operation. FM 2-0 and FM 2-01.3 provide information on how to conduct IPB

6-12. One of the most significant contributions that intelligence personnel can accomplish is to accurately predict future enemy COAs. Predictive intelligence enables the commander and staff to anticipate enemy actions and develop corresponding plans or counteractions. Commanders must receive intelligence in time to make an effective decision, create orders for subordinate units, and have them act on it. Commanders develop and rely on their CCIR to focus collection of information supporting decision points (DP) in the commander's plan.

6-13. The process works best when the BCT staff, or the staffs of subordinate battalions, anticipates CCIR well in advance. A proactive BCT S-2 will anticipate the commander's priority intelligence requirements to help drive the planning of reconnaissance operations. The S-2's requirement management techniques focus the collection, processing, and intelligence production on the critical needs of the commander. The intelligence staff assists the commander with battlefield visualization by identifying feasible threat capabilities; confirming or refuting the threat's COAs; and providing accurate descriptions of the effects of the operational environment on friendly and threat activities. In the BCT, tactical operations center (TOC), the command and control (C2) digital systems enable a near-continuous assessment of the operations cycle. The S-2's role in IPB is a continuous activity to obtain information and produce intelligence essential to the commander's decision-making.

# INTELLIGENCE RUNNING ESTIMATE

6-14. After developing threat COAs, the S-2 and MI company, supported by the rest of the BCT staff, develop the intelligence running estimate. IPB products make up the basis of the intelligence estimate. The intelligence running estimate:

- Forms the basis for the facts and assumptions of the MDMP, driving various staff section running estimates and the remaining steps in the MDMP.
- Is a logical and orderly analysis of the terrain, weather, and civil considerations of the operational environment and its effects on friendly and threat COAs, threat capabilities and vulnerabilities, threat tactics, techniques and procedures (TTP), and the probability of adoption of threat COAs.
- Provides the best possible answer to the commander's PIRs that are available at the time.
- Is dynamic and changes constantly with the situation.

6-15. The S-2 briefs the intelligence running estimate results to the brigade commander and staff. Upon conclusion of the staff briefings, the commander states his intent for the operation, and provides additional planning guidance to the staff. The commander's guidance to the S-2 could include:

- Additional threat COAs and objectives to consider.
- Additions or deletions of threat decision points and high-value targets (HVT).
- Approval or modification of recommended PIRs.
- Specific instructions on priority for and allocation of reconnaissance and surveillance assets.

## SECTION III – SUPPORT TO RECONNAISSANCE OPERATIONS

# SYNCHRONIZATION

6-16. ISR synchronization is the task that:

- Analyzes information requirements and intelligence gaps.
- Evaluates available assets (internal and external).
- Determines gaps in the use of those assets.
- Recommends reconnaissance assets controlled by the organization to collect on CCIR, and submits requests for information for adjacent and higher collection support.

6-17. The intelligence officer synchronizes the collection effort in coordination with the operations officer, MI company officers, and other staff elements as required. This effort includes recommending tasks for assets that the commander controls, and submitting requests for information to adjacent and higher echelon units and organizations. See FM 2-0 for detailed information on the intelligence synchronization process.

6-18. The many sources of information available to the BCT must be synchronized with requirements to collect relevant information, and then produce the intelligence the BCT requires. The result is a continuous feed of information that enables the commander to maintain SA and make timely decisions. The BCT S-3 integrates the sensors and other capabilities of the BCT to accomplish this. The reconnaissance squadron's role is to obtain information for the BCT commander, answering his CCIR. This information gathering also assists the BCT S-2 in confirming the intelligence estimate. Information from all the organic, adjacent, higher and joint, interagency, intergovernmental, and multinational (JIIM) collection assets are integrated to form the common operational picture (COP). The BCT S-2 takes this information, analyzes it, and gives the BCT commander refined intelligence products with which to make decisions. The MI company supports the BCT S-2 in synchronization, planning, and data analysis. MI company assets, such as unmanned aircraft systems (UAS), Prophets, and human intelligence (HUMINT) collection teams (HCT), may be task organized to subordinate battalions or other units within the BCT.

# INTEGRATION

6-19. ISR integration is the task of assigning and controlling a unit's reconnaissance assets (in terms of space, time, and purpose) to collect and report information as a concerted and integrated portion of operation plans and orders. This task ensures assignment of appropriate assets through a deliberate and coordinated effort integrating reconnaissance activities into the operation.

## INTEGRATED RECONNAISSANCE OPERATIONS

6-20. Reconnaissance operations are integrated between higher, lower, and adjacent units to ensure that collection is complete but not redundant. ISR plans are collaborative efforts that support CCIR. Planning includes coordinating reconnaissance handover from the higher headquarters (HQ) to the BCT, and from the BCT to battalions (both reconnaissance squadron and maneuver). During planning and liaison with the division, the BCT ensures that the width and depth of their assigned reconnaissance AO are suitable for its reconnaissance and surveillance capabilities. The BCT develops named areas of interest (NAI) to assist in focusing assets, either to discern enemy courses of action and disposition of forces, or to gain information about the environment. Planners must ensure that plans are synchronized by taking into account the battle rhythms of subordinate, adjacent, supporting, and higher headquarters. ISR working groups at each organization must synchronize their own requirements and schedules with others to avoid a duplication of effort, and to ensure there are not gaps in surveillance.

## SECTION IV – RECONNAISSANCE CONSIDERATIONS

# COMMANDER'S RECONNAISSANCE DECISIONS

6-21. The BCT commander will be required to make several decisions relating to reconnaissance and surveillance (including MI discipline collection). Based on the results of those decisions, different COAs may be selected, including development of a new mission, a continuation of the current mission, and/or the anticipation of transition into a new mission. Generally, these decisions are covered in the BCT standard operating procedure (SOP), and usually are issued in the commander's planning guidance. These decisions include:

- Commander's intent for intelligence collection.
- CCIR.
- The degree of collaborative reconnaissance and surveillance planning between the BCT and reconnaissance squadron staffs.
- When to issue the ISR plan and with which order (warning order [WARNO] #2 or operation order [OPORD]).
- When to rest and refit reconnaissance assets.

## COLLABORATIVE INTELLIGENCE, SURVEILLANCE, AND RECONNAISSANCE PLANNING

6-22. The BCT staff collaborates with the reconnaissance squadron in developing the BCT ISR plan. The BCT commander decides to what degree he wants his primary collector, the reconnaissance squadron, involved in BCT-level planning. He has three basic choices:

- If reconnaissance squadron and BCT command posts (CP) are collocated or close to each other, then squadron personnel can be involved easily.
- If the reconnaissance squadron is executing an operation and its CP is located far from the BCT main CP, member(s) of the reconnaissance squadron staff could move to the BCT main CP to assist in planning and serve as a liaison officer(s) (LNO).
- The reconnaissance squadron staff may participate collaboratively through a variety of FM radio or Army Battle Command System (ABCS) methods.

## ISSUING THE INTELLIGENCE, SURVEILLANCE, AND RECONNAISSANCE PLAN

6-23. The commander should decide when and how he wants to issue the ISR plan. After developing his CCIR, the commander must decide how soon he wants reconnaissance assets to start collecting, and how much (or little) synchronized planning he is willing to accept before reconnaissance execution. Factors that affect this decision include current status/capability of the BCT collection assets; availability of non-BCT collection assets to collect BCT IR; and level of integration required with the overall brigade scheme of maneuver.

6-24. The commander should issue the ISR plan at the earliest occasion following completion of mission analysis. This enables the BCT to begin the reconnaissance effort early to help refine planning for the BCT scheme of maneuver. Other options include the following:

- **After COA approval.** The commander delays the reconnaissance effort to ensure synchronization with the scheme of maneuver. The commander uses this method when time constraints do not require reconnaissance assets to be positioned early during planning.
- **Totally separate from a BCT order.** The commander often uses this method during continuous operations or after the higher HQ issues a WARNO for a follow-on mission. If the commander issues the ISR plan prior to finalizing the BCT OPORD, it is more likely the ISR plan will require fragmentary orders (FRAGO) to support the refined maneuver plan.

## RESTING AND REFITTING RECONNAISSANCE AND SURVEILLANCE ASSETS

6-25. The third decision the commander must make is when to rest and refit his reconnaissance and surveillance assets. The commander must think this decision through early and build it into the overall concept of operations. The BCT's ISR plan must account for available troops to task at a given time. Coordination with higher HQ is critical for the timing of the BCT battle rhythm as requests for alternate reconnaissance and surveillance elements to provide necessary BCT coverage take time to coordinate. As another option, the commander may have redundant means of coverage within the BCT to conduct a reconnaissance handover with his primary collectors.

# COMMANDER'S RECONNAISSANCE GUIDANCE

6-26. Reconnaissance must begin as early as possible, with more direct and early involvement by the commander to focus his staff and subordinate units during the process. Guidance should be pushed forward early and refined later. The BCT's MDMP begins with the receipt of a new mission. Usually this new mission is part of the initial WARNO from the division. The initial WARNO sent by the BCT should provide the following information for reconnaissance planning to begin:

- Division mission and concept of the operation.
- Division commander's intent.
- BCT reconnaissance objective.
- CCIR and other IR.
- Focus, tempo, and engagement criteria.
- Specified reconnaissance tasks.

# INTELLIGENCE, SURVEILLANCE, AND RECONNAISSANCE WORKING GROUP

6-27. The ISR working group is a temporary grouping of designated staff representatives who coordinate and integrate intelligence collection, and reconnaissance and surveillance activity, and provide recommendations to the commander. The ISR working group usually includes:

- BCT executive officer (chairs the meeting).
- BCT S-3 (alternate chair) or representative.
  - Engineer coordinator (ENCOORD) representative.
  - Air defense and airspace management (ADAM)/brigade aviation element (BAE) representative.
- BCT S-2 or representative.
- MI company commander/collection manager or representative.
- Reconnaissance squadron S-3 and/or S-2 or representative.
- S-2X or representative.
- Brigade fire support officer or representative.
- BCT S-7 or representative.
- Command, control, communications, and computers cell representative.
- BCT S-9 or representative.
- CBRN officer.
- Sustainment cell representative.
- Subordinate unit representatives (if available).
- Special operations forces (SOF) representative (if available).

6-28. The ISR working group applies ISR synchronization into other BCT processes. It allocates reconnaissance assets to gather information for CCIR; verifies and updates the ISR tasking matrix; and ensures support by each warfighting function for each reconnaissance asset. The ISR working group meeting is a critical event that must be integrated effectively into the BCT's battle rhythm to ensure the collection effort provides focus to operations, rather than disrupting them. Preparation and focus are essential to a successful working group. Each representative must come to the meeting prepared to discuss available assets, capabilities, limitations, and requirements related to his warfighting function. Planning the working group's battle rhythm is paramount to conducting effective reconnaissance operations. The working group cycle should be scheduled to complement the higher headquarters' battle rhythm and its subsequent requirements and timelines.

6-29. The BCT S-3 (or representative) must be prepared to provide the following information:

- Current friendly situation.
- Reconnaissance assets available.
- Requirements from higher HQ (including recent FRAGOs or taskings).
- Changes to the commander's intent.
- Changes to the task organization.
- Planned operations.

6-30. The BCT S-2 (or representative) must be prepared to provide the following:

- Current enemy situation.
- Current CCIR.
- Current ISR plan.
- Situational template tailored to the time period discussed.
- Collection assets available and those the S-2 must request from higher HQ.
- Weather and effects of weather on intelligence collection.

## RECONNAISSANCE PLANNING CONSIDERATIONS

6-31. The ISR working group must consider several factors when developing the ISR plan:

- The requirement for reconnaissance assets in follow-on missions (e.g., reconnaissance squadron mission during close fight).
- The amount of time BCT planners will have to refine the ISR plan during the execution of current operations.
- Modifications to the MDMP that might overlap ISR planning with COA development.
- Risk the commander is willing to accept if reconnaissance activity is begun before the ISR plan is fully integrated with the scheme of maneuver.
- Use of dismounted reconnaissance assets to include insertion methods, movement once committed, and extraction methods.
- Locations of command and control nodes (e.g., proximity of the BCT command group and the reconnaissance squadron CP).
- Locations of fire support (FS) assets (e.g., positioned to support extended reconnaissance elements).
- Reconnaissance handover between the BCT and higher HQ assets, or between the reconnaissance squadron and maneuver battalions.
- Sustainment support to include casualty evacuation (CASEVAC), resupply, and vehicle recovery.
- Counterreconnaissance plan.

# INTELLIGENCE, SURVEILLANCE, AND RECONNAISSANCE PLAN ATTACHMENTS

## INTELLIGENCE, SURVEILLANCE, AND RECONNAISSANCE OVERLAY

6-32. The ISR overlay expresses the ISR plan in graphic form Subordinate reconnaissance units may determine some of the control measures instead of the BCT. If so, the BCT must consolidate the control measures into the overlay as soon as the unit completes its planning. If the overlay is transmitted over digital systems it might need to be broken into component parts to speed transmission and reduce clutter.

For example, it could be broken into one overlay showing the basic operational graphics and boundaries, one showing infiltration graphics, and one showing sensor locations and range fans. See FM 2-0 for more information on the development of the ISR overlay.

### ENEMY SITUATION TEMPLATE

6-33. The S-2 develops an enemy situation template for the operation that focuses on the enemy's reconnaissance and counterreconnaissance efforts. The S-2 designs the enemy situation template to aid in planning friendly infiltration and survivability by identifying enemy actions that will impact friendly reconnaissance efforts. It also includes enemy main body activities. This information keeps the reconnaissance unit focused on the reconnaissance objective.

### INTELLIGENCE, SURVEILLANCE, AND RECONNAISSANCE SYNCHRONIZATION MATRIX

6-34. The ISR synchronization matrix is used to plan and direct the collection effort. It may consist of a list of available collection means, plus brief notes or reminders on current intelligence requirements and specific information to collect. The S-2 section initially prepares the ISR synchronization matrix, which the ISR working group completes, and the S-3 authorizes as part of the ISR plan. See FM 2-0 and FMI 2-01 for information about development of the ISR synchronization matrix.

6-35. It is critical that as the BCT's scheme of maneuver becomes refined (usually after wargaming) and the planners reevaluate the ISR plan to determine if the plan needs any modification. The ISR plan tasks assets or organizations across the brigade that conducts reconnaissance and surveillance. The S-3 maintains a current status of reconnaissance assets' availability, capability, vulnerability, and performance. This assessment, coupled with the CCIR and time requirements focused on DP, drives the ISR plan.

6-36. The BCT S-3 consolidates ISR overlays from higher and adjacent headquarters, and subordinate reconnaissance, combined arms, and fires battalions, along with any attached or supporting battalion-sized external elements. He may also require smaller elements to submit reconnaissance or surveillance plans, and sustainment battalions as well. The staff must quickly ensure both that the various ISR plans are synchronized (for gaps and overlaps), and that the ISR plan and the BCT order remain synchronized. The BCT commander and staff must be prepared to make adjustment decisions for reconnaissance assets to account for the developing situation.

6-37. The BCT S-3, with S-2 and MI company support, assesses ISR integration and synchronization continuously, monitoring for variances in actual coverage of NAI and targeted areas of interest (TAI) from the plan, and looking for requirements for additional coverage. Anticipated DP are closely monitored using the indicators for the CCIR supporting them, and CCIR are updated as some become irrelevant and new ones are added for the developing situation.

## RECONNAISSANCE IN STABILITY OPERATIONS

6-38. One of the purposes of reconnaissance during stability operations is to develop the intelligence needed to address the issues driving an insurgency. Several factors are particularly important for reconnaissance operations in stability operations. These include the following:

- A focus on the local populace.
- Collection occurring at all echelons.
- Localized nature of insurgencies.
- All Soldiers functioning as potential collectors.
- Insurgent use of complex terrain.

6-39. Affecting intelligence synchronization is the requirement to work closely with U.S. Government agencies, host nation security and intelligence organizations, and multinational intelligence organizations. Operational-level reconnaissance planning drives the synchronization of these agencies' and organizations' efforts; however, coordination occurs at tactical echelons. Communication among collection managers and collectors down to the battalion level is important. It can eliminate circular reporting and unnecessary duplicate work. External intelligence staffs and leaders may require integration into the BCT's tactical network to enable closer coordination.

6-40. Counterinsurgency (COIN) operations may require reorganization of BCT intelligence personnel. Each company may require a company intelligence support team, which would include analysts. A reconnaissance element is also essential. Linguists are a battle-winning asset, but like any other scarce resource, commanders must allocate them carefully. During pre-deployment, the best use of linguists may be to train Soldiers in basic language skills. BCT commanders and staffs may have to integrate contracted civilian linguists into their operations and planning.

6-41. A BCT optimized for COIN operations may require political and cultural advisors. The current force structure gives corps and division commanders a political advisor, and brigade and battalion commanders an information engagement officer or cell. Lower echelon commanders must improvise. BCT leaders may use culturally aware Soldiers organic to the unit to perform these functions. The position requires someone with "people skills and a feel" for the environment. They help the commander shape the environment by planning and assessing information engagement tasks.

# SURVEILLANCE

6-42. The BCT integrates surveillance, to include higher and joint surveillance capabilities, into its ISR plan. Conducting surveillance is systematically observing the airspace, surface, or subsurface areas, places, persons, or things in the AO by visual, aural, electronic, photographic, or other means. Surveillance activities include:

- Orienting the surveillance asset on the NAI and/or the surveillance objective in a timely manner.
- Reporting all information rapidly and accurately.
- Completing the surveillance mission no later than the time specified in the order.
- Answering the requirement that prompted the surveillance task.

6-43. Surveillance is distinct from reconnaissance. Often surveillance is passive and may be continuous. Reconnaissance missions are typically shorter and use active means (such as maneuver). Additionally, reconnaissance may involve fighting for information. Reconnaissance involves many tactics, techniques, and procedures throughout the course of a mission. An extended period of surveillance may be one of these. Commanders complement surveillance with frequent reconnaissance. Surveillance, in turn, increases the efficiency of reconnaissance by focusing those missions while reducing the risk to Soldiers. The BCT tasks surveillance responsibilities in the same manner as reconnaissance missions, using the ISR overlay and tasking matrix.

6-44. National and joint wide-area and focused surveillance missions can provide valuable information to the BCT. While national and joint surveillance systems focus on information requirements for combatant commanders, they also provide information to all commanders in support of operations across the area of operations. The systematic observation of geographic locations, persons, networks, or equipment is assigned to Army intelligence, reconnaissance, and maneuver assets. Changes or anomalies detected during surveillance missions can generate a reconnaissance mission to confirm or deny the change.

## SECTION V – FORMS OF RECONNAISSANCE

6-45. There are four forms of reconnaissance: route, zone, area, and reconnaissance in force. The BCT commander uses one of these forms to logically group specific information requirements and taskings into missions for subordinate commanders. The four forms of reconnaissance refine the scope of the commander's mission further and give it a spatial relationship. The BCT, as part of division operations, may be assigned reconnaissance missions. The CCIR and specified tasks dictated by higher HQ usually refine the associated tasks listed below. The commander must think through the tasks that he wants his reconnaissance and surveillance assets to conduct, and ensure that he task organizes appropriately. For instance, if he wants them to conduct clearance of obstacles, reconnaissance elements might require engineers and additional maneuver forces (because obstacles are generally overwatched by an enemy force).

## ROUTE RECONNAISSANCE

6-46. A route reconnaissance is an operation focused on obtaining detailed information on a specific route and all adjacent terrain from which the threat could influence the route. The route may be a road or trail or other linear feature. Units conduct route reconnaissance to ensure the route is clear of obstacles and threat, and that it will support planned movement. A route reconnaissance may be performed as part of an area or zone reconnaissance. Typically, route reconnaissance missions are assigned to organizations below battalion level, although a BCT could be assigned the primary mission of route reconnaissance of a main route to be used by a larger force. As a brigade mission, the BCT may task organize similar to a movement to contact mission (see Chapter 2).

## ZONE RECONNAISSANCE

6-47. Zone reconnaissance is the directed effort to obtain detailed information concerning threat, terrain, society, and infrastructure in accordance with the commander's reconnaissance focus within a location delineated by boundaries (i.e., a line of departure, lateral boundaries, and a limit of advance). A zone reconnaissance is assigned when the enemy situation is vague or when information concerning cross-country trafficability is desired. It is appropriate when previous knowledge of the terrain is limited or when combat operations have altered the terrain. The reconnaissance may be threat-oriented, terrain-oriented, society-oriented, infrastructure-oriented, or a combination. Additionally, the brigade commander may focus the reconnaissance effort on a specific force, such as the enemy's reserve. A terrain-focused zone reconnaissance must include the identification of obstacles, both existing and reinforcing, as well as areas of chemical, biological, radiological, and nuclear (CBRN) contamination. See FM 3-90 for the conduct of a zone reconnaissance.

6-48. When higher headquarters assigns the BCT a zone reconnaissance, it may, based on its mission analysis and identification of the subsequent specified and implied tasks, perform a combination of the forms of reconnaissance to answer the higher commander's information requirements. Therefore, even though the brigade as a whole is performing a zone reconnaissance mission, its subordinate units could be assigned a zone, area, or route reconnaissance mission.

6-49. A zone reconnaissance is organized with subordinate elements operating abreast of one another within a portion of the zone as designated by graphic control measures. If the BCT commander expects to find significant enemy forces within the zone, he considers attaching armored, mechanized, or aviation forces to the reconnaissance elements to deal with the anticipated threat. If it is likely that reconnaissance elements will encounter significant obstacles or mobility impediments, he may also augment them with combat engineers. While reconnaissance missions are the specialty of the reconnaissance squadron, all BCT combat units may perform this mission. The HBCT commander may use three balanced task forces operating in adjacent zones, each with reconnaissance, tank, Infantry, and artillery company-sized units. Unless augmented with wheeled transportation, IBCT commanders have limited capability to conduct brigade-size zone reconnaissance except in very complex terrain with little depth that they can conduct on foot.

## AREA RECONNAISSANCE

6-50. An area reconnaissance is a directed effort to obtain detailed information concerning the terrain or threat activity within a prescribed area. The BCT conducts area reconnaissance by maneuvering elements through the area, or by establishing observation posts (OP) within or external to the area. The BCT may conduct area reconnaissance as a primary mission for a higher headquarters, or may assign the mission to its subordinate elements, in particular the reconnaissance squadron. See FM 3-90 for the conduct of area reconnaissance.

6-51. When higher headquarters assigns the BCT an area reconnaissance, it may, based on its mission analysis and the subsequent identification of specified and implied tasks, perform a combination of the forms of reconnaissance to answer the higher commander's information requirements. Its subordinate battalion task forces could be assigned a zone, area, or route reconnaissance mission.

6-52. Forces conducting an area reconnaissance are organized according to the size, geography, physical infrastructure, and social dynamics of the area to be reconnoitered; the time available for conducting the reconnaissance; and the capabilities of the organization performing the reconnaissance. The BCT may be required to reconnoiter one large area or several smaller ones. In many cases, areas to be reconnoitered are given to subordinate elements. Area reconnaissance proceeds faster than zone reconnaissance because the effort is focused on a relatively smaller, specific piece of terrain or threat force. Mission analysis will determine the task organization of various subordinate elements required to accomplish their specific reconnaissance responsibilities. In many cases, JIIM elements (e.g., civil affairs teams) will be attachments to reconnaissance forces.

## RECONNAISSANCE IN FORCE

6-53. A BCT conducts a reconnaissance in force to discover or test enemy strengths, disposition, and reaction capability or to obtain other information. It is conducted when the enemy is known to be operating within an area and adequate intelligence cannot be obtained by other means. It is an aggressive reconnaissance, conducted as an offensive operation in pursuit of clearly stated CCIR. It differs from other reconnaissance operations because it is normally conducted only to gain information about the enemy and not the terrain, and the risk of engagement with the enemy is expected. See FM 3-90 for discussion on the conduct of a reconnaissance in force mission.

6-54. The BCT commander task organizes his subordinate units based upon mission analysis. If heavy resistance is expected, a combined arms task force with an attached reconnaissance troop may act as the main effort. Less resistance and more reconnaissance objectives may require a reconnaissance squadron main effort with a tank or Infantry company attached. In complex terrain or against fortified defenses, Infantry-heavy task forces may be required.

This page intentionally left blank.

# Chapter 7

# Fire Support in the Brigade Combat Team

Fire support of the Brigade Combat Team (BCT) includes lethal and nonlethal fires that generate the commander's desired effects. Lethal fire support comes from the BCT's organic indirect fires assets, Army artillery and aviation assets, and joint and multinational artillery and aviation assets. Nonlethal fire support can come from a wide range of military and civilian, joint and multinational partners. The BCT commander and staff must exercise command and control over a diverse group of assets from a wide range of sources in ways that will successfully integrate and synchronize them with BCT operations. The BCT fires cell works closely with the fires battalion on recommendations to the BCT commander on the command or support relationships of any additional field artillery units supplementing the BCT.

## SECTION I – BRIGADE COMBAT TEAM FIRE SUPPORT ORGANIZATION

## FIRES CELL

7-1. The BCT fires cell and its elements integrate the fires warfighting function in BCT operations. It is led by the brigade fire support officer and staffed by Soldiers who have expertise that is integral to the fires warfighting function. It has resources to plan for future operations from the main command post (CP) and to support current operations from the tactical command post (TAC CP) (when deployed). Additionally the cell has the limited capability to provide coverage to the command group and the deputy command group when deployed. Fires cell staff are assigned to the following elements within the fires cell of the main CP and current operations cell of the TAC CP:

- Lethal fires element (main CP).
- Nonlethal fires element (main CP).
- Air support element (Air Force tactical air control party ). The air support element operates from the main CP with selected portions deployed with the TAC CP when the TAC CP is deployed.
- TAC CP fires element. Selected personnel from the lethal and nonlethal fires elements deploy with the TAC CP when the TAC CP is deployed, otherwise they are part of main CP fires cell.

7-2. All elements work from the main CP if the TAC CP is not deployed (Figure 7-1). Selected personnel from the main CP's lethal and nonlethal fires elements man the fires element of the TAC CP when the TAC CP is deployed. The rest remain at the main CP. The BCT Air Force tactical air control party (TACP) collocates with the fires cell in the main CP, but is sufficiently robust that a selected portion of it can be deployed with the TAC CP. As mission variables (mission, enemy, terrain and weather, troops and support available, time available, and civil considerations [METT-TC]) dictate, the fires cell in the main CP can be augmented by other Army, Air Force, or joint resources and assets including those for information engagement, civil affairs and related activities as needed.

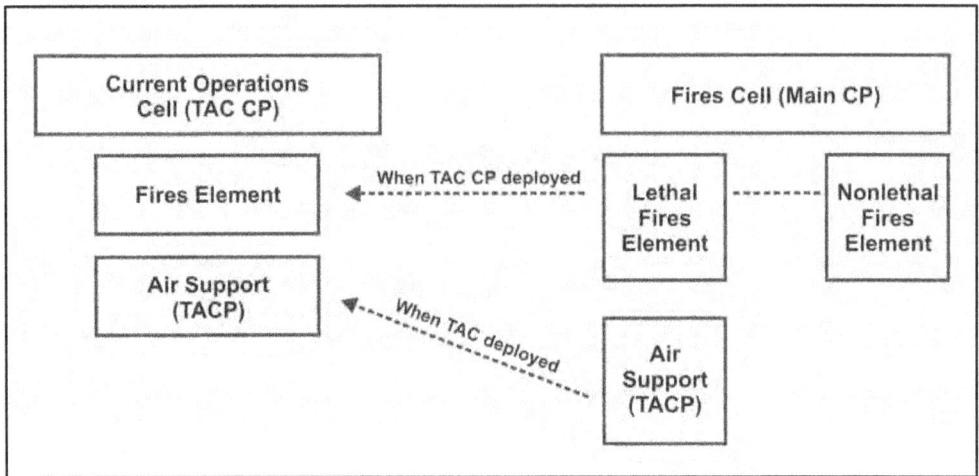

**Figure 7-1. BCT fires cell**

7-3. The fires cell is the centerpiece of the BCT's targeting architecture, focused on both lethal and nonlethal target sets. The fires cell thus collaboratively plans, coordinates and synchronizes fire support, aspects of information engagement (such as artillery and air delivered leaflets) and civil affairs in an integrated fashion with the other warfighting functions to support BCT operations. The targeting working group brings together representatives of all staff sections concerned with targeting. It synchronizes the contributions of the entire staff to the work of the fires cell. The brigade operational law team is co-located with the fires cell in order to provide legal review of plans, targeting and orders. The fires cell coordinates and integrates joint fire support into the BCT commander's concept of operations. Primary fires cell functions include:

- Planning, coordinating, and synchronizing fire support for BCT operations.
- Working with the BCT staff to integrate lethal and nonlethal fires, including appropriate aspects of information engagement and civil affairs operations, into the BCT targeting process.
- Collaborating in the intelligence preparation of the battlefield (IPB) process.
- Coordinating the tasking of sensors during development of the intelligence, surveillance, and reconnaissance (ISR) plan with the BCT S-2, the military intelligence company commander (as needed), and the reconnaissance squadron to acquire targets.
- Participating in the BCT's military decision-making process.
- Briefing the BCT commander to get his approval of the concept for fire support.
- Disseminating the approved concept to BCT fire support organizations, the fires battalion, the division's fires brigade, and the division and corps fires cells.
- Participating in the BCT targeting process.
- Ensuring battalion fires cells plan fires in accordance with the BCT commanders guidance for current and future operations.
- Preparing the fires paragraphs in the BCT operation order (OPORD) that describe the concept/scheme of fires to support BCT operations.
- Managing the establishment of, and changes to, fire support coordination measures.
- Coordinating maneuver space for the positioning of field artillery assets.
- Coordinating clearance of lethal and nonlethal attack against targets (clearance of fires).
- Coordinating assessment of effects generated by fires.
- Coordinating requests for additional fire support to include joint fires.
- Providing input to the BCT's common operational picture to enhance situational understanding.

# INFORMATION OPERATIONS CELL

7-4. The information operations (IO) cell focuses on information engagement, one of the Army's information tasks. It also helps synchronize the other elements of Army information tasks with the rest of the staff. These include:

- Command and control warfare (electronic warfare [EW]) with the brigade fire support officer, brigade electronic warfare officer, and S-6.
- Information protection (information assurance and computer network defense) with the S-6.
- Operations security (including physical security of information systems) with the S-2 and S-3.
- Military deception with the S-2 and S-3.

7-5. The IO cell is anchored on the S-7, S-9, and public affairs officers who plan, coordinate, integrate, and synchronize all aspects of information engagement and civil affairs activities to support BCT operations. The IO cell assists the brigade fire support officer to integrate information engagement capabilities into the targeting process. They are part of the targeting working group that integrates information engagement and civil affairs operations into the targeting process. The fires cells at battalion level and fires support teams at company level, to include organic mortars and any additional fire support allocated by the BCT fires cell, for delivery on time and on target. Together the BCT fires cell and the fires cells of subordinate BCT organizations enable the BCT to control fires that protect the force and shape the battlefield. See Chapter 8 for more information on the conduct of information engagement and civil-military operations in support of BCT operations.

7-6. Planning information engagement and civil affairs operations requires integrating them with several other processes including IPB and targeting. The executive officer (XO) synchronizes these activities within the overall operation. The S-7, S-9, public affairs officer and brigade judge advocate develop supporting tasks during course of action development and finalize them during course of action analysis. During planning these tasks are discussed in terms of information engagement or civil-military operations (CMO) tasks, measures, or activities.

# TACTICAL AIR CONTROL PARTY

7-7. An Air Force tactical air control party (TACP) is collocated with the BCT headquarters (HQ). The TACP is resourced to support both the BCT tactical command post and the BCT main command post. The air liaison officer (ALO) advises the BCT commander and staff on air operations. The ALO leverages the expertise of the BCT TACP with links to the division TACP to plan, coordinate, synchronize, and execute air support operations. He also maintains situational awareness of the total air support and air support effects picture.

7-8. TACP functions include:

- Serving as the Air Force commander's representative providing advice to the BCT commander and staff on the capabilities, limitations, and employment of air support, airlift, and reconnaissance.
- Providing an Air Force coordination interface with the BCT fires section, air defense and airspace management (ADAM)/brigade aviation element (BAE) section, and maneuver battalions.
- Coordinating activities through the Joint Air Request Net (JARN) and the advanced airlift notification net.
- Helping to synchronize air and surface fires and prepare the air support plan.
- Providing direct liaison for local ADAM activities.
- Integrating into the staff to facilitate planning air support for future operations, and providing advice about the development and evaluation of close air support (CAS), interdiction, reconnaissance, and joint suppression of enemy air defenses (JSEAD) programs.
- Providing terminal control for CAS and operating the Air Force air request net.

## JOINT FIRES STAFF AUGMENTATION

7-9. Joint augmentation is essential to BCT operations. In addition to the Air Force TACP, other joint augmentation may include Naval surface fire support (NSFS) and U.S. Marine Corps (USMC) air or artillery liaison officers.

## COMBAT OBSERVATION AND LASING TEAMS

7-10. The combat observation and lasing team (COLT) is an observer team controlled at the brigade level, capable of day and night target acquisition and that has both laser-range finding and laser-designating capabilities. The brigade fire support officer is responsible for training the COLTs and for performing precombat checks and mission briefings/rehearsals before employment. The BCT fires cell supervises the planning and execution of COLT employment. COLTs can be used as independent observers to weight key or vulnerable areas. Although originally conceived to interface with the Copperhead, a COLT can be used with any munitions that require reflected laser energy for final ballistic guidance. The self-location and target ranging capabilities of the Fire Support Sensor System (FS3) mounted with the M1200 Armored Knight, which replaces the M707 Knight vehicle with the ground/vehicular laser locator designator or the FS3, enables first-round fire for effect with conventional munitions.

## FIRES BATTALION

7-11. The BCT's organic fires battalion provides field artillery fires to the BCT and subordinate units in priority, and for missions that may be assigned by the BCT commander. The fires battalion also provides reactive counterfire against enemy mortar, cannon, and rocket elements in the BCT's area of operations. FM 3-09.21 provides a description of the field artillery battalion (Figure 7-2). The BCT's fires battalion has an established organic command relationship. However, the BCT commander may assign a support relationship of direct support, reinforcing, general support, or general support-reinforcing. A BCT's fires battalion should be prepared to assume any assigned mission. This may include missions not normally given to a field artillery unit (for example, base defense, patrolling, search and rescue, or flood relief). This is most likely to occur during stability or civil support operations. Stability operations are discussed in greater detail in Chapter 4 of this manual and in FM 3-07.

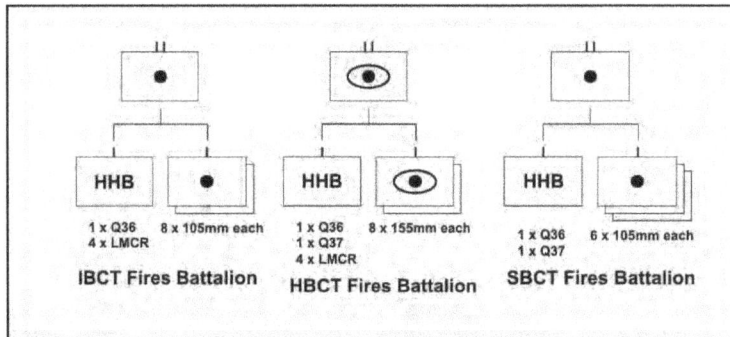

Figure 7-2. Fires battalions

## SECTION II – FIRE SUPPORT PLANNING

7-12. Fire support planning follows the operations process described in FM 3-0. FM 6-20 (to be revised, renumbered and renamed) provides a detailed overview of the fire support planning process. FM 6-20-40 and FM 6-20-50 include descriptions of BCT tactics, techniques, and procedures for fire support planning and coordination.

7-13. Targeting is the process of selecting and prioritizing targets and matching the appropriate response to them, considering operation requirements and capabilities (Joint Publication [JP] 3-0). Targeting is effective when its results support the commander's objectives. Units identify lethal and nonlethal targeting options based on those objectives. Fire support planning is integrated with and supports the targeting process described in FM 6-20-10 (to be revised and renumered). The targeting methodology process follows the sequence of decide, detect, deliver, and assess.

# DECIDE, DETECT, DELIVER, AND ASSESS PROCESS

## DECIDE

7-14. The BCT commander and his entire staff play a significant role in the "decide" function. The decide function provides the overall focus, identifies targeting requirements, and sets the initial priorities and planning for specific reconnaissance activities. These activities support the "detect" function; the attack methodology that supports the "deliver" function; and command and control (C2), battle damage assessment (BDA), and intelligence issues in the "assess" function. As part of the decide function, the BCT should answer the following questions:

- What targets should be acquired and attacked?
- When and where are the targets likely to be found?
- How long will the target remain once acquired?
- Who or what can locate the targets?
- What accuracy of target location will be required to attack the target?
- What are the priorities for reconnaissance objectives and asset allocation?
- What priority intelligence requirements (PIR) are necessary, and how and when (no later than time/date) must the information be obtained, processed, and disseminated?
- What is the risk to the environment or local cultural and historical resources?
- When, where, how, and in what priority should the targets be attacked?
- What are the effects criteria that must be achieved to attack the target successfully?
- Who or what can attack the targets, and how should the attack be conducted (for example, number/type of attack elements, ammunition) to maximize effects and resources based on commander's guidance?
- What or who will obtain the BDA or other information the BCT needs to determine the success or failure of each attack? Who must receive and process that information, how rapidly, and in what format?
- Who has the decision-making authority to determine success or failure, and how rapidly must the decision be made and disseminated?
- What actions will be required if an attack is unsuccessful and who has the authority to direct those actions?

7-15. The BCT staff prepares several products as they work through the decide process. These products must complement the commander's scheme of maneuver and provide the basis for the concept of fires. Typical fire support (FS) products include:

- **High-value target list (HVTL).** The HVTL is a list of targets or assets essential for the enemy commander to accomplish his mission. The loss of high-value targets (HVT) would be expected to degrade important enemy functions seriously throughout the friendly commander's area of interest. The brigade fire support officer and S-7 identify HVTs during mission analysis and course of action (COA) development.
- **High-payoff target list (HPTL).** The HPTL is a by-phase, prioritized list of those HVTs that must be acquired and successfully attacked for the success of the friendly commander's mission. Examples of high-payoff targets (HPT) are the enemy's C2 nodes and intelligence collection systems. The HPTL is a dynamic document that is continually refined during both planning and execution based on the situation and the commander's guidance. Usually, the HPTL is identified through wargaming (Table 7-1).

### Table 7-1. Example of high payoff target list

| Phase of the Operation: I – Isolate the Enemy Units | | |
|---|---|---|
| Priority | Category | HPT |
| 1 | Fire support | Insurgent mortars |
| 2 | Maneuver | Insurgent teams |
| 3 | Command and control | Insurgent cell phone |
| 4 | Command and control | Insurgent FM radio |

- **Target selection standards (TSS).** This matrix focuses on accuracy to establish criteria for deciding when targets are located accurately enough to attack. Military intelligence (MI) analysts use TSS to determine targets from combat information and pass them to FS elements for attack (Table 7-2).

### Table 7-2. Example of target selection standards

| HPT | Timeliness | Accuracy |
|---|---|---|
| Insurgent mortars | 10 minutes | 100 meters |
| Insurgent teams | 30 minutes | 100 meters |
| Insurgent cell phone | Within 2 hours of H-hour | Placed/received within 12 km of Fustina airfield |
| Insurgent FM radio | 20 minutes | 150 meters |

- **Attack guidance matrix.** The commander must approve this matrix, which addresses the targets or target sets to attack, how and when they will be attacked, and the desired effects that attacking the target will generate (Table 7-3).

### Table 7-3. Example of attack guidance matrix

| HPT | When | How | Effect | Remarks |
|---|---|---|---|---|
| Insurgent mortars | I | FA | D | Use search and attack teams in restricted areas. |
| Insurgent teams | I | FA | N | Destroy C2. |
| Insurgent cell phone | A | EA | EW | Disrupt service starting H-2. |
| Insurgent FM radio | A | EA | EW | No jamming until H-3 to preserve intelligence. |
| Legend | | | | |
| When:<br>I = Immediate<br>A = As acquired<br>P = Planned | How:<br>FA = Field Artillery<br>EA = Electronic Attack | | Effects:<br>S = Suppress<br>N = Neutralize<br>D = Destroy<br>EW (electronic warfare) = Jamming | |

- **Target synchronization matrix (TSM).** The TSM combines data from the HPTL, the ISR plan, and the attack guidance matrix (AGM). It lists HPT by category and the units responsible for detecting them, attacking them, and assessing the effects of the attacks.
- **Sensor/attack matrix.** This matrix is a targeting tool used to determine whether the critical HVT can be acquired and attacked. The matrix enables war game participants to record their assessments of the ability of sensor systems to acquire and attack HVT at a critical event or phase of the battle.
- **Combat assessment requirements.** The requirements for combat assessment are identified during COA development. Combat assessment consists of BDA, munitions effectiveness assessment, and re-attacks recommendations.

- **Target nominations.** These may include targets nominated for attack by higher HQ. These include air interdiction, Army Tactical Missile Systems (ATACMS), and electronic attack.

7-16. Other BCT sections work with the brigade fire support officer and S-7 to prepare products that complement fires and IO products. These products include:

- ISR plan. The ISR Plan focuses primarily on answering CCIRs and then identifying HPT. The ISR plan is prepared by the S-3 and coordinated with the S-2. It is a major contributor to the detect and assess functions.
- Decision support template (DST). The DST is a tool planners use to anticipate and synchronize required friendly actions at critical junctures on the battlefield. It is one method the BCT uses to tie target execution to the friendly scheme of maneuver.

## Effects to be Generated by Attacking the Target

7-17. Based on the BCT commander's guidance, the targeting working group recommends how each target should be engaged in terms of the degree and duration of desired effects. The desired effects generated by attacking the target should be clearly identified. Effect can be a standard term such as harass, suppress, neutralize, or destroy; or it may be described in further detail such as the length of the effect (FM 6-20-10). (See Table 7-4.)

### Table 7-4. Examples of fire effects

| Effect | FM 1-02 Definition | Example |
|--------|--------------------|---------|
| Harass | Fires designed to disturb the rest of enemy troops, to curtail movement and, by the threat of losses, to lower morale. | Random fires on enemy defensive positions during periods of reduced visibility. |
| Suppress | Fires on or around a weapons system to degrade its performance below the level needed to fulfill its mission objectives. | Fire directed at enemy air defense systems to reduce threat to unmanned aircraft systems. |
| Neutralize | Fires to render the target ineffective or unusable for a temporary period. | Preventing the enemy from using a bridge for vehicle traffic for 4 hours. |
| Destroy | Fires to physically render the target permanently noncombat-effective or so damaged that it cannot function unless it is restored, reconstituted, or rebuilt. | Killing 60% of the personnel in an enemy Infantry battalion and seriously wounding another 20% of the personnel. |

7-18. Fires planners must understand the differences between broad targeting objectives that support the commander's intent, and the detailed targeting objectives and effects determination required for the execution of specific fire plans and programs. An overall objective of disrupting an enemy's tempo (commander's goal) by blocking an enemy force for X-hours or until a certain time (tactical level effect) can be achieved through the use of:

- Fires to interdict the enemy's route (tactical effect)—
  - Fires to deliver scatterable mines (SCATMINE) at chokepoints; and fires to cause avalanches (1st order direct physical effect) that block (2nd order indirect functional effect) or obstruct routes.
  - Fires to damage bridges or crater highways (1st order direct physical effect) or otherwise cut lines of communications that render a movement route unusable (2nd order indirect functional effect).
- Fires against a unit that disrupt/slow the unit's movement (tactical effect). Destructive fires against engineer, bridging, or fuel equipment (1st order direct physical effect).
- Destructive fires against lead vehicles at chokepoints or key C2 vehicles (1st order direct physical effect). Harassing fires (tactical task) that could create fear, havoc, and complications (2nd order indirect behavioral effect) by destroying and damaging equipment, killing and

wounding personnel, obscuring vision (1st order direct physical effect); and that could cause the enemy to travel buttoned-up, at open column (2nd order indirect behavioral effect).

## DETECT

7-19. The S-3, assisted by the S-2, is responsible for directing the ISR plan to detect HPT identified in the decide function. Execution of detect functions must be timely and accurate. Based on the results and effectiveness of the execution of detect functions, the products developed in the decide function may be modified.

7-20. The detect function involves locating HPT accurately enough to engage them. Characteristics and signatures of the relevant targets are determined and then compared to potential attack system requirements to establish specific sensor requirements. The S-2 section works closely with the fires section to identify the specific "who, what, when, and how" for target acquisition. Information needed for target detection is expressed as PIRs and/or information requirements (IR) to support the attack of HPT and associated fire support tasks. As target acquisition assets gather information, they report their findings back to the commander and staff. Detection plans, priorities, and allocations change during execution based on METT-TC. As part of the detect function, the BCT should answer the following questions:

- Were the designated targets found at the anticipated locations, times, and conditions, and to the required accuracies?
- Are detect-related plans, units, and equipment performing as required? Are there any combat loss or maintenance issues? Is detect information being processed and disseminated in a timely manner?
- Is the situation developing as anticipated (i.e., the threat characteristics main effort identification, friendly/enemy success/failure)?
- Have reconnaissance and surveillance activities identified new, unanticipated information that must be considered?
- Based on detect functions, are changes required to other decide, detect, deliver, and assess (D3A) functions?

7-21. The information gathered from the multitude of collection assets must be processed to produce targets meeting TSS. Moving HPT must be detected and tracked to maintain a current target location. The brigade fire support officer and S-7 planners tell the S-2 the accuracy required and dwell time for a target to be eligible for attack. To facilitate the hand-off of targets and tracking (as in reconnaissance handover), the S-3 coordinates with higher and subordinate units to establish responsibilities.

7-22. The fires section assists the S-2 in the detect function by providing information from the field artillery (FA) radars and observers to help complete the intelligence picture. Critical targets not attacked must be tracked to ensure they are not lost. Tracking suspected targets expedites execution of the attack guidance and keeps the targets in view while they are validated. The fires section monitors fire missions and spot reports (SPOTREP) for targeting information.

## DELIVER

7-23. The deliver function of the targeting process is based on the attack guidance and the selection of an attack system or combination of systems. As part of the deliver function, the BCT should answer the following questions:

- Can/should designated targets be attacked as planned or are changes required?
- Are the established attack guidance and effects criteria still valid and achievable?
- Are lethal and nonlethal delivery systems performing as required? Are there any combat loss or maintenance issues?
- Are unanticipated delivery requirements manageable, or are there actual or potential implications?

- Based on deliver factors (i.e., ammunition or weapon status), are changes required to other D3A functions?
- What effect will this have on the civilian infrastructure, civilian population, environment and cultural and historical resources? What level of collateral damage is acceptable and who is authorized to make that decision?

7-24. Successful attack of HPT requires the brigade fire support officer and S-7 to:

- Determine if the planned attack system is available and is still the best system for attack. Coordinate as necessary to use the attack system in a timely manner. (e.g., use of offensive electronic warfare systems requires coordination with the electronic warfare officer, the signal officer, the IO officer, and the S-2).
- Deconflict and synchronize all attacks as necessary to gain maximum, synergistic effects with minimum resource expenditure.
- Ensure IO and EW, especially the destructive component, are properly incorporated into the overall targeting plan.
- Coordinate as required with higher, lower, and adjacent units, other services, allies, and the host nation . This is particularly necessary to minimize the risk of fratricide.
- Issue the call for fire to the appropriate executing unit(s).
- Inform the S-2 of the target attack.

7-25. The deliver function involves engaging targets located within the TSS according to the guidance in the AGM. This includes using both lethal and nonlethal attack systems. HPT that are located within the TSS are tracked and engaged at the time designated in the OPORD/AGM. Other collection assets look at HPT that are not located accurately enough or for targets within priority target sets. When one of these is located within the TSS, its location is sent to the system that the AGM assigns to attack it. Not all HPT will be identified accurately enough to be attacked before execution. Some target sets might not have very many targets identified. Collection assets and the intelligence system develop information that locates or describes potential targets accurately enough to engage them. The HPTL sets the priority in which they are attacked. The attack of targets requires a number of tactical and technical decisions and actions. These decisions determine the:

- Time of attack.
- Desired effects.
- Attack system (s) or IO activity to be used.

7-26. Based on these decisions, technical decisions can be made using the following considerations as an outline:

- Delivery means (lethal and nonlethal).
- Number and types of munitions or systems, tools, and techniques.
- Unit to conduct the lethal/nonlethal attack.
- Response time of the asset or unit to provide the effects.

## ASSESS

7-27. The commander and staff assess the results of mission execution. Assessment occurs throughout the operations process. Targets are attacked until the effects outlined in the AGM are achieved or until the target is no longer within the TSS. If combat assessment reveals that the commander's guidance has not been met, the detect and deliver functions of the targeting process must continue to focus on the targets involved and make adjustments to the plan as necessary.

7-28. Assessment recommendations could result in changes to original decisions made during the decide function. These changes must be provided to subordinate units as appropriate because they impact continued execution of the plan. In the case of IO effects, the time it takes to achieve desired effects, or impact the target or audience, might be lengthy. The assessment of these effects cannot be treated and processed as rapidly as a lethal effect would be.

7-29. The brigade fire support officer, S-7, and targeting working group must consider the following when conducting combat assessment:

- Impacts on achieving the commander's intent if targets supporting a fire support task were engaged but the desired effects or objectives were not achieved.
- Whether or not BDA and measures of effectiveness can be objective enough to measure the achievement of the commander's intent.
- The degree of accuracy of the assessment relies largely upon collection resources and their quality as well as quantity.

7-30. Battle damage assessment begins with a micro-level examination of the damage or effect resulting from attack of a specific target, and ends with macro-level conclusions regarding the functional outcomes created in the target system. BDA is conducted in three phases:

- The first phase examines the outcomes at the specific targeted elements.
- The second phase estimates the functional consequences for the target system components.
- The third phase projects results on the overall functioning of the target system and the consequent changes in the enemy's behavior.

*Future Targeting and Re-Attack Nominations*

7-31. The final phase of combat assessment is nominating future targets or re-attacks. In this phase, by closely examining what was done (battle damage assessment), and how it was done (munitions effectiveness assessment), a determination can be made as to whether the desired effects have been generated by attacking this particular target with the selected means. If the desired effect was not achieved, a determination must be made as to whether the same target should be re-attacked using the same means or a different, perhaps more effective means. It might be possible that the desired effect simply cannot be generated by re-attacking this particular target, and therefore, an entirely new target or set of targets must be attacked to achieve the desired effect. This last activity both completes the targeting process and begins it anew by linking effects actually generated with those desired at the beginning of the targeting cycle.

7-32. The effects of nonlethal attacks require continuous assessment. The S-7 is responsible for this assessment and monitors reporting based on information requirements and requests for information from higher headquarters. The S-7 uses the measures of effectiveness established during COA analysis to maintain a continuous assessment. Based on this assessment, the S-7 decides whether to continue to engage the target, to break off the attack, or to engage the target with another IO element. The S-7 bases his decision on the extent to which continuing to engage the target will increase the likelihood of accomplishing the IO objectives it supports, and the extent to which accomplishing the IO objectives will contribute to completing the mission.

# FIND, FIX, FINISH, EXPLOIT, ANALYZE, AND DISSEMINATE

7-33. FM 6-20-10 introduces a methodology for conducting non-lethal targeting: find, fix, finish, exploit, analyze, and disseminate (F3EAD). F3EAD is a sub-process of D3A. It is a means of integrating lethal and nonlethal elements into the planning cycle of the military decision-making process (MDMP) to ensure that every effort is directed toward achieving the commander's desired effect. F3EAD provides the maneuver commander an additional tool to address certain targeting challenges, particularly those found in a counterinsurgency environment:

- **Find.** The target is identified and the target's network is mapped and analyzed.
- **Fix.** A specific location and time to engage the target are identified and the validity of the target is confirmed.
- **Finish.** This mirrors the deliver function of D3A when the action planned against the target is initiated and completed. The finish step differs from the deliver function in D3A by the nature of the means the commander will apply against identified target sets.
- **Exploit.** The engaging unit takes the opportunity to gather additional information.
- **Analyze.** The engaging unit determines the implications and relevance of the information.
- **Disseminate.** The engaging unit publishes the results.

# TARGETING MEETINGS

## PURPOSE OF TARGETING MEETINGS

7-34. Targeting meetings integrate the targeting process into other BCT processes. The purpose of a targeting meeting is to focus and synchronize the unit's combat power and resources toward finding, attacking, and assessing current HPT by using the D3A methodology. The meeting verifies and updates the HPTL; verifies, updates, and modifies tasking of the collection assets for each HPT; allocates delivery systems to engage each target; and confirms the assets tasked to assess the effects on the target(s) after the attack. A successful targeting meeting requires preparation by participants, participation by all warfighting function representatives, and the rapid development and dissemination of required products. Specific objectives of targeting meetings include:

- Verification and update of the HPTL.
- Verification, update of, and re-tasking available collection assets for each HPT.
- Allocation of delivery systems to engage each target.
- Confirmation that assets are tasked to assess whether the desired effects have been achieved by engaging the target.
- Identification of target nominations for attack by division or joint assets.
- Synchronization of lethal and nonlethal actions (including IO).
- Synchronization of FS and IO assets to generate desired lethal and nonlethal effects.

## TARGETING WORKING GROUP

7-35. The targeting working group is a grouping of predetermined staff representatives concerned with targeting who meet to provide analysis, coordinate and synchronize the targeting process, and provide recommendations to the targeting board. The targeting working group usually includes the staff primaries or representatives of the following:

- Brigade fire support officer (leads and plans/coordinates the working group).
- BCT S-3 (alternate lead).
- BCT S-2 representative.
- BCT S-4 Sustainment cell representative.
- BCT S-6 representative.
- BCT S-7 representative.
- BCT S-9 representative.
- Engineer representative.
- ADAM/BAE representative.
- Electronic warfare representative.
- Fires section targeting officers.
- Tactical air control party representative.
- Military information support operations (MISO) representative.
- Civil affairs (CA) unit representative.
- Brigade judge advocate representative.
- Fires battalion S-3 and S-2 representatives.
- Reinforcing field artillery liaison officer (LNO).
- Naval surface fire support LNO.
- Brigade provost marshal (PM), crime analyst, or military police representative.

## TARGETING BOARD

7-36. The targeting board is a temporary grouping of designated predetermined staff representatives with decision authority to coordinate and synchronize the targeting process. The targeting board usually includes:

- BCT XO (chairs the board).
- BCT S-3 (alternate chair).
- BCT S-2.
- BCT S-6.
- BCT S-7.
- BCT S-9.
- Fire support coordinator.
- Brigade fire support officer.
- Engineer coordinator (ENCOORD).
- ADAM/BAE officer.
- Electronic Warfare officer.
- Air liaison officer.
- Fires cell targeting officers.
- MISO team leader.
- Civil affairs unit leader.
- Brigade judge advocate.
- Sustainment cell representative.
- Fires battalion S-3 and S-2.
- Reconnaissance squadron S-3, S-2, and fire support officer (FSO).
- Maneuver battalion S-3, S-2, and FSO.
- Brigade special troops battalion (BSTB) and brigade support battalion (BSB) fire support noncommissioned officers (FSNCO).
- Military intelligence company (MICO) commander/collection manager.
- Special operations forces (SOF) representative.
- Reinforcing field artillery unit liaison officer.
- Naval surface fire support LNO.

## TARGETING RESPONSIBILITIES

### Brigade Commander

7-37. The BCT commander's intent focuses and drives the targeting process. He approves the recommendations of the targeting working group.

### Executive Officer

7-38. The BCT XO usually chairs the targeting board. Although the BCT commander must approve the initial targeting products that accompany an operation plan (OPLAN)/OPORD, the XO (or deputy commanding officer) may be the approval authority for modifications to targeting products.

### Brigade Combat Team Fire Support Officer

7-39. The brigade fire support officer finalizes the fires attack guidance formulated by the commander. His specific targeting responsibilities include:

- Leading the targeting working group in preparation for the targeting board.
- Overseeing fires targeting execution.

- Ensuring all aspects of targeting are addressed and understood during the targeting process (i.e., task, purpose, location of sensor/back-up, fire mission thread, rehearsal, delivery asset, and assessment).
- Developing and updating the fire support tasks.
- Consolidating target refinements and planned targets from subordinate units.
- Establishing target refinement standards to facilitate completion of the FS plan.
- Coordinating support for subordinate unit attack requirements.
- Coordinating suppression of enemy air defenses (SEAD) and joint air attack team (JAAT).
- Assessing Phase I BDA to determine if the desired effects were achieved.
- Formulating the re-attack recommendation.
- Ensuring target nominations are validated, processed, and updated to support the air tasking order (ATO).

## Brigade Combat Team S-2

7-40. The S-2 must work in concert with the entire staff to identify collection requirements and implement the ISR plan. The S-2 determines collection requirements, develops the ISR matrix with input from the staff representatives, and continues to work with the staff planners to develop the ISR plan. The S-2 also identifies those reconnaissance and surveillance (to include MI discipline collection) assets which can provide answers to the commander's targeting requirements. The S-2's specific targeting responsibilities include:

- Providing weather effects on targeting operations.
- Providing information about enemy capabilities and projected courses of action.
- Providing IPB products to the targeting team.
- Developing HVTs.
- Determining which HPT can be acquired with organic assets.
- Developing support requests for acquiring HPTs beyond the capabilities of organic assets.
- Coordinating the collection and dissemination of targeting information with the fires section.
- Advising the S-3 about BDA collection capabilities.

## Brigade Combat Team S-3

7-41. The BCT S-3's specific targeting responsibilities include:

- Working with the S-2, S-7, and brigade fire support officer to prioritize the HPTL before approval by the commander.
- Developing and supervising implementation of the ISR plan.
- Determining the targets to be attacked immediately and desired effects.
- Providing a detailed interpretation of the commander's concept of the operation.
- Providing guidance about which targets are most important to the commander.
- Deciding when and where targets should be attacked.
- Periodically reassessing the HPTL, AGM, and BDA requirements with the brigade fire support officer, S-7, and S-2.
- Determining with the brigade fire support officer, S-7, and S-2 if desired effects have been achieved and if additional attacks are required.

## Brigade Combat Team S-7

7-42. The S-7 is responsible for the overall planning, preparation, execution and assessment of information tasks for the BCT. His targeting responsibilities include:

- Synchronizing appropriate aspects of information engagement with the fires, maneuver and other warfighting functions.
- Assessing enemy vulnerabilities, friendly capabilities, and friendly missions.

- Nominating information engagement-related targets for attack.
- Briefing deception operations.
- Providing operation security measures.
- Synchronizing Army information tasks.

## Brigade Combat Team S-9

7-43. The S-9's targeting responsibilities include:
- Providing advice on the affects of friendly actions on the civilian populace.
- Producing input to the restricted target list.
- Providing assessments of the effectiveness of CA activities.
- Brigade judge advocate.
- The brigade judge advocate's targeting responsibilities include:
  - Providing advice on rules of engagement impacts on targeting.
  - Providing advice on law of war impacts on targeting.

## Brigade Judge Advocate

7-44. The brigade judge advocate's targeting responsibilities include:
- Providing advice on rules of engagement impacts on targeting.
- Providing advice on law of war impacts on targeting.

## Targeting Officer

7-45. The targeting officers in the fires section facilitate the exchange of information between the BCT S-2 and subordinate fires sections. Their responsibilities include:
- Helping the BCT S-3/S-2 develop ISR plans.
- Developing, recommending, and disseminating the AGM.
- Coordinating with the BCT S-2 for target acquisition (TA) coverage and processing of HPT.
- Producing the TSS matrix for target acquisition assets supporting the BCT.
- Managing target lists for planned fires.
- Coordinating and distributing the restricted target list in coordination with division.

## Air Liaison Officer

7-46. The ALO's targeting actions include:
- Monitoring execution of the ATO.
- Advising the commander and staff about employment of air assets.
- Receiving, coordinating, planning, prioritizing, and synchronizing immediate CAS requests.
- Providing Air Force input to analysis and plans.

## ADAM/BAE Officer

7-47. The ADAM/BAE officer's targeting actions include:
- Providing brigade airspace requirements and airspace control measures.
- Deconflicting airspace user requirements with targeting requirements.
- Providing the air defense plan.
- Providing the air picture and early warning plans.

## Electronic Warfare Officer

7-48. The electronic warfare officer's targeting responsibilities include:

- Determining HPT to engage with electronic attack.
- Recommending electronic warfare methods of target engagement.
- Planning and coordinating tasking and requests to satisfy electronic attack requirements.
- Assisting the S-2 with the electronic portion of IPB.
- Identifying threat electronic attack capabilities and targets.

## Senior Military Information Support Operations Non-Commissioned Officer

7-49. The MISO noncommissioned officers' (NCO) targeting responsibilities include:

- Identifying targets with MISO relevance and recommending their placement or removal from the targeting list during the nomination process and selection board.
- Coordinating MISO targeting with deception.
- Coordinating with supporting MISO units.
- Synchronizing BCT MISO with higher and lateral HQ.
- Providing assessments of the effectiveness of MISO activities.

## Engineer Coordinator

7-50. The ENCOORD's targeting responsibilities include:

- Providing advice on the use of scatterable mines.
- Providing input to the restricted target list (including relevant environmental considerations).
- Providing advice on the obstacle creating effects of indirect fires.
- Effects of terrain as it relates to targeting.

## Brigade Provost Marshal

7-51. The provost marshal's targeting responsibilities include:

- Providing advice on crime trends, patterns, and associations.
- Coordinating and preparing warrants for target folders and packages.
- Tracking and providing detainee disposition and reporting.

## FREQUENCY AND TIMING OF TARGETING MEETINGS

7-52. The targeting meeting is a critical event in the BCT's battle rhythm. It must be integrated effectively into the BCT's battle rhythm and nested within the higher HQ targeting cycle to ensure the results of the targeting process focus, rather than disrupt operations. Thus, task organization changes, modifications to the ISR plan, ATO nominations, and changes to the HPTL all must be made with full awareness of time available to prepare and execute.

7-53. The timing of the targeting meeting is critical. While the timeline for brigade level targeting meetings is usually 24 to 36 hours out, higher HQ assets and certain targeting decisions, such as ATO nominations usually are based on a 72-hour cycle.

7-54. Experience in the BCT has shown the benefits of two targeting meetings daily at the main CP. A preliminary meeting (the targeting working group) facilitated by the fire support officer ensures the targets and objectives of nonlethal effects complement those of lethal effects, and that both meet the commander's guidance and intent. The brigade fire support officer and S-7 assess ongoing targeting efforts and ensure target nominations to higher HQ are being processed in a timely manner. The second meeting (the targeting board) is generally more formal than the first, is chaired by the XO, and has as its focus updating the commander and gaining new guidance and approval of planned and proposed targeting actions in the next 24, 48 and 72 hours. Adequate time should be allowed for targeting meetings to present targeting information and situation updates, and to provide recommendations and obtain decisions. Experience has shown that about 60 minutes is the norm. Those BCTs conducting two targeting meetings assign the responsibility for nonlethal targeting to the deputy commanding officer (DCO), while the BCT XO retains responsibility for lethal targeting.

## PREPARING FOR THE TARGETING MEETING

7-55. Preparation and focus are keys to successful BCT targeting meetings. Each representative must come to the meeting prepared to discuss available assets, capabilities, limitations and BDA requirements related to their staff area. This means participants must conduct detailed prior coordination and be prepared to provide input and/or information. This preparation must be focused around the commander's intent and a solid understanding of the current situation.

7-56. The BCT S-3 must be prepared to provide:

- Current friendly situation.
- Maneuver assets available.
- Current combat power.
- Requirements from higher HQ (including recent fragmentary orders or tasking).
- Changes to the commander's intent.
- Changes to the task organization.
- Planned operations.
- Current ISR plan.

7-57. The BCT S-2 must be prepared to provide:

- Current enemy situation.
- Planned 3nemy COAs (situation template) tailored to the time period discussed.
- Collection assets available and those the S-2 must request from higher HQ.
- Weather and weather effects on operations.

7-58. The brigade fire support officer must be prepared to provide:

- Changes to the fire support tasks.
- Fire support assets available.
- Proposed HPTL, TSS, AGM, and fire support tasks for the time period discussed.
- Recommended changes to fire support coordination measures (FSCM) for period being discussed.
- Any changes to ammunition control or supply rates.

7-59. The S-7 must be prepared to provide:

- Changes to the fire support tasks.
- IO assets available.
- Proposed HPTL, TSS, AGM, and fire support tasks for IO for the time period discussed.

7-60. The specific situation dictates the extent of the remaining targeting team members' preparation. They should be prepared to discuss in detail (within their own warfighting function) available assets and capabilities, the integration of their assets into targeting decisions, and the capabilities and limitations of enemy assets. The following tools should be available to facilitate the conduct of the targeting meeting: HPTL, TSS, AGM, consolidated matrix (e.g., target synchronization matrix) or another product per standard operating procedure (SOP), list of delivery assets, and list of collection assets.

7-61. The following tools should be available to facilitate targeting meetings:

- **TSM.** The TSM visually illustrates the HPTs and is designed to list specific targets with locations in each category. The matrix has entries to identify whether a target is covered by a named area of interest (NAI); the specific detection, delivery, and assessment assets for each target; and attack guidance. Once completed, the TSM serves as a basis for updating the AGM and issuing a fragmentary order (FRAGO) at the conclusion of the meeting. In addition, it facilitates the distribution of the target meeting's results.

- **Situation Template (SITEMP).** The SITEMP identifies where specific targets are expected to be found on the battlefield and ensures assets are properly placed to detect them.
- **List of potential detection and delivery assets.** A list of all potential detection and delivery assets available to the unit helps all attendees visualize which assets might be available for detection and delivery. It is essential that staff members be prepared to discuss the potential contribution for the particular assets within their area.

## SECTION III – FIRE SUPPORT COORDINATION

7-62. Fire support coordination ensures the synchronization of fire support assets to match the right attack means with the correct target to deliver the BCT commander's desired effects at the precise time and location needed to support BCT operations. To achieve the best possible synchronization of all fire support, particularly in joint operations, the following guidelines for coordination are recommended:

- Position indirect fire weapons systems and units to support the commander's concept of operations.
- Coordinate use of naval surface fire support and planned and immediate CAS to achieve the BCT commander's intent and concept of operations.
- Ensure that the brigade fire support officer, subordinate fire support officers and observers know the exact locations of maneuver boundaries and other fire support coordination measures.
- Position observers in redundancy where they can see their assigned targets and trigger points, communicate with fire support assets, and respond to the maneuver commander.
- Establish field artillery and mortar final protective fires and priority targets. Final protective fire is an immediately available prearranged barrier of fire designed to impede enemy movement across defensive lines or areas.
- Plan field artillery illumination to facilitate direct fire during limited visibility.
- Provide common survey to mortars.
- Provide meteorological (met) data to mortars.
- Use the fire support execution matrix to execute fire support and remain flexible to branches or sequels to the current plan.
- Coordinate with the fires battalion tactical operations center to develop the AGM using the munitions effects database in the Advanced Field Artillery Tactical Data System (AFATDS). Compute ammunition requirements needed for generating desired effects via the attack of expected enemy target categories with fire support. Provide this assessment to the BCT commander so that he can formulate his attack guidance. Also, compute ammunition requirements and identify issues that require the BCT commander's attention or additional guidance, such as fire support tasks that may be unsupportable.
- Disseminate target priorities to the BCT staff and through to the lowest levels of subordinate maneuver units, fire support organizations and mortars.
- Develop and disseminate field artillery-delivered SCATMINES safety boxes in coordination with the BCT engineer coordinator and the S-3.
- State the BCT commander's attack guidance by defining how, when, and with what restrictions the commander wants to attack different targets and identify the targeting priorities. The data should be entered into the AFATDS database.
- Require refinement by lower fire support echelons to be completed by an established cut-off time.
- Verify or correct target locations and trigger points during refinement.
- Recommend the risk the BCT commander is willing to accept concerning delivery of indirect fires for maneuver units in close contact. Calculate risk estimate distances.
- Consider limiting the number of targets to 10 to 15 per maneuver battalion/squadron going down, with no more than 45 to 60 for the entire BCT.
- Use the fire support execution matrix to brief the fire support portion of the OPORD during the combined arms rehearsal.

- During the combined arms rehearsal, rehearse the fire support portion of the OPORD directly from the fire support execution matrix.
- Conduct rehearsals with the actual Soldiers who will execute fire support tasks (e.g., the forward observer who will initiate fires on a critical target rather than his company/troop fire support officer).
- Ensure methods for battle tracking and clearance of indirect fires are clearly understood by fires cells and maneuver commanders.
- Verify the range of Q-36, Q-37 and Q-48 radar, and field artillery and mortar delivery system coverage based on the effects desired and appropriate shell/fuse combinations.
- Prioritize requirements for Q-36, Q-37 and Q-48 radar and allocate radar zones to reflect the developed SITEMP, protection priorities and the scheme of maneuver. Ensure radar zones are within the coverage and trajectory arc of the radar systems. For details, see FM 3-09.21.
- Explain fire support-related combat power in terms of the required effects to be generated for the operation. The BCT and subordinate maneuver commanders then better understand fire support contributions to the course of action scheme of maneuver. Useful information may also include:
  - Number and type of missions available/possible.
  - Battery/battalion/mortar volleys available by the type of ammunition and the effects expected.
  - Minutes of obscurants available and allocation.
  - Minutes of illumination available and allocation.
  - Number of available SCATMINE by type, size, density, and safety zone.

## CLEARANCE OF FIRES AND INFORMATION TASKS

7-63. Maneuver commanders clear fires. At the BCT level, the commander usually delegates the responsibility for coordinating fires to the brigade fire support officer. The BCT accomplishes clearance of fires in any of several ways:

- Through a staff process.
- Through control measures embedded in automated battle command systems.
- Through active or passive recognition systems.

7-64. Clearance of fires ensures fires will attack enemy capabilities without resulting in casualties to friendly forces and noncombatants. Similarly, clearance of effects ensures that any electronic attack conducted by the BCT does not interfere with friendly or civilian operations. Even with automated systems, clearance of fire support remains a command responsibility at every level, and commanders must assess the risk and decide the extent of reliance on automated systems.

7-65. The brigade fire support officer and S-7 coordinate all fire support impacting in the BCT AO and information tasks affecting the BCT. They ensure that fire support and the execution of information tasks will not jeopardize troop safety, will interface with other FS and information tasks, and/or will not disrupt adjacent unit operations.

### RISK

7-66. The level of risk is directly related to the level of situational awareness. The BCT commander determines the acceptable risk level when delegating clearance of fires to subordinate units. Units under fire need immediate responsive FS and cannot wait for a multi-layered clearance process through higher HQ to receive indirect fire.

## SUPPRESSION OF ENEMY AIR DEFENSES

7-67. The effective employment of air assets gives the BCT commander a powerful source of fires. Aviation assets enable the ground commander to influence operations quickly and to add depth to the battlefield. The suppression of enemy air defenses allows the BCT to use these assets to its maximum advantage. FM 3-01.4 provides additional detailed information.

# Chapter 8

# Augmenting Combat Power

The Brigade Combat Team (BCT) commander augments the BCT's maneuver units with lethal and nonlethal resources and units based on his assessment of the mission variables and the BCT's concept of operations (e.g., prioritizes the main effort). Sections I through VIII discuss how these lethal and nonlethal assets protect and support the BCT. Sections X and XI of this chapter provide a discussion of the sources of external augmentation that could be made available to the BCT (e.g., close air support [CAS], naval gunfire).

Within the Infantry Brigade Combat Team (IBCT) and Heavy Brigade Combat Team (HBCT), the brigade special troops battalion (BSTB) contains the assigned BCT elements that protect and/or support maneuver units. The Stryker Brigade Combat Team (SBCT) does not have a BSTB. It controls its supporting units under its brigade headquarters (HQ). These units consist of the engineer company, the military intelligence company (MICO), the brigade signal company, the military police (MP) platoon, and the chemical, biological, radiological, and nuclear (CBRN) reconnaissance platoon. In the HBCT and IBCT the engineer company is organic to the BSTB. In the SBCT, the engineer company is a separate company reporting to the BCT HQ.

The BCT can expect to receive augmentation upon arrival in a new theater. These augmenting units may come from a support brigade such as a sustainment or maneuver enhancement brigade (MEB). For instance, an explosive ordnance disposal (EOD) company will come from an EOD group. Based on the factors of mission, enemy, terrain and weather, troops and support available, time available and civil considerations (METT-TC), and the commander's guidance, the BCT staff integrates these assets into maneuver operations and organizations at all levels.

## SECTION I – ENGINEERING OPERATIONS

8-1.  Combat engineering performs essential mobility, countermobility, and survivability tasks for the BCT as they support assured mobility, enhancing protection, enabling expeditionary logistics and support to building capacity. It includes capabilities organic to and augmenting the BCT. It may be augmented at times with general engineering support, but retains its focus on the integrated application of engineer capabilities to support the combined arms unit's freedom of maneuver and protection. See FM 3-34.22 for more detailed information.

## ENGINEER STAFF

8-2.  The engineer staff within the BCT HQ includes the engineer coordinator (ENCOORD), an engineer section, and a terrain team. The staff engineer section synchronizes BCT engineer planning and integrates it into the BCT military decision-making process (MDMP). It operates as a part of the assured mobility section with other mobility elements, providing support throughout the BCT area of operations (AO). The ENCOORD uses the essential mobility survivability tasks format to communicate the BCT commander's priorities for his available engineering assets to subordinate units (FM 3-34). The staff engineer section is capable of preparing executable engineer plans and orders that require minimal refinement by subordinate units. This capability, coupled with digital dissemination of information, minimizes the need for time-consuming engineering planning at battalion and company levels.

8-3.   The staff engineer section task organizes and performs staff supervision for organic and augmented engineer forces and any host nation, coalition, or contracted engineer support under BCT control. The section digitally tracks, reports, analyzes, and disseminates all engineer and terrain-related information that might influence BCT operations including an explosive hazards database. An obstacle database includes all confirmed obstacles, mines, munitions, and unexploded ordnance (UXO) encountered by the force during any action or operation. The section conducts tracking and database management in conjunction with EOD elements operating within the BCT AO.

8-4.   The terrain team is the focal point for geospatial information and products within the BCT. Using its organic Digital Topographic Support System (DTSS), the terrain team provides the BCT with timely digital terrain products and integrated terrain analysis. The team also enables the BCT to obtain other geospatial products through reach back capabilities. The team provides the commander with a clear understanding of the physical environment by enabling visualization of the terrain and explaining its impact on friendly and enemy operations.

# ENGINEER COMPANIES

8-5.   Organic engineer companies in each of the BCTs provide assured mobility, enhance protection, and enable logistics and increased engineer capacity. Their primary focus is on assured mobility for maneuver elements. BCT engineer companies have very limited countermobility and survivability capabilities. They rely heavily on the integration of scatterable mine systems and complex terrain to support temporary defensive actions. BCTs require augmentation by engineer forces to support longer duration defensive actions (FM 3-34 and FM 3-34.22). Each engineer company is organized slightly differently from any other (Figure 8-1).

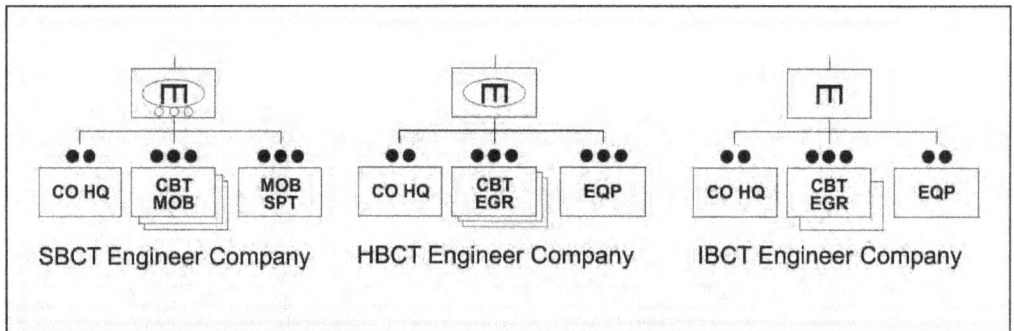

**Figure 8-1. BCT engineer companies**

## SECTION II – MILITARY POLICE OPERATIONS

8-6.   The BCT provost marshal (PM) serves as a special staff officer to the BCT commander, and provides guidance and direction to the BCT MP platoon leader regarding MP operations. The PM is responsible for planning, coordinating, and synchronizing MP assets and functions. The MP platoon organic to a BCT consists of a platoon HQ and three MP squads. Each squad has four three-man teams and four vehicles (two light trucks and two "Guardian" armored security vehicles). Higher HQ may provide additional MP assets to augment each BCT. Depending upon METT-TC, the brigade could receive additional MP assets ranging from platoon- to company-size units with relevant specialties (e.g., military working dog teams, investigators, traffic, physical security).

# MILITARY POLICE PLATOON CAPABILITIES

8-7.  These organic MP platoons provide the minimum essential MP capabilities to support BCT operations. The MP platoon is capable of performing prioritized tasks from within any of the MP functions. MPs enable the BCT commander to achieve their objectives by providing a unique set of functional capabilities. These capabilities support all joint functions and the Army warfighting functions. MPs accomplish this through their own functions of law and order, police intelligence operations, internment and resettlement, maneuver and mobility support, and area security. See FM 3-39 for additional information.

## LAW AND ORDER

8-8.  MPs work to reduce the opportunity for criminal activity throughout an AO by assessing the local conditions; conducting police engagement at all levels to include coordinating and maintaining liaison with other DOD, host nation, joint, and multinational agencies; and developing coherent policing strategies. MP units at all levels coordinate actions to identify and influence crime conducive conditions that might promote random and organized criminal activity, or that have the potential to threaten a tactical line of effort or a Soldier. MPs support and develop strategies to maintain order and enforce the rule of law across the spectrum of operations.

8-9.  Close coordination with host nation civilian police can enhance MP efforts at combating terrorism, maintaining law and order, and controlling civilian populations. The law and order function also includes major areas such as law enforcement, criminal investigations, host nation police training and support, and support to U.S. Customs operations.

## POLICE INTELLIGENCE OPERATIONS

8-10.  Police intelligence operations is a military police function, integrated within all MP operations. Police intelligence operations support MP operations through analysis, production, and dissemination of information. This information was collected through police activities conducted to enhance situational understanding, protection, civil control, and law enforcement (FM 3-39). This information, whether police, criminal, or tactical in nature, is gathered during MP operations. Upon analysis, the information may contribute to the commander's critical information requirements (CCIR); intelligence-led, time-sensitive operations; or policing strategies needed to forecast, anticipate, and preempt crime or related disruptive activities to maintain order.

## INTERNMENT AND RESETTLEMENT OPERATIONS

8-11.  Military police conduct internment and resettlement operations to shelter, sustain, guard, protect, and account for populations (detainees, dislocated civilians, or United States military prisoners) as a result of military or civil conflict, natural or manmade disaster, or to facilitate criminal prosecution. Internment involves detaining a population or group who pose some level of threat to military operations. Resettlement involves the quartering of a population or group for their protection. These operations inherently control the movement and activities of their specific population for imperative reasons of security, safety, or intelligence gathering (FM 3-39.40).

## MANEUVER AND MOBILITY SUPPORT

8-12.  Maneuver and mobility support is a military police function conducted to support and preserve the commander's freedom of movement and enhance the movement of friendly resources in all environments (FM 3-39). The maneuver and mobility support operation aids and enhances the maneuver commander's freedom of movement and maneuver. MP units expedite the secure movement of theater resources to ensure that commanders receive forces, supplies, and equipment needed to support the operational plan and changing tactical situations.

AREA SECURITY

8-13. Military police (MP) assist the brigade commander in addressing security and protection. The goal is to enhance maneuver unit freedom, enabling the unit to conduct missions without placing unnecessary requirements on BCT maneuver forces. Area security actions include reconnaissance, surveillance, antiterrorism measures, and security of designated personnel, equipment, facilities, and critical points. These actions also include convoy and route security.

# ADDITIONAL MILITARY POLICE SUPPORT TO THE BCT

8-14. Any one of the five MP battlefield functions easily could require more than one MP platoon. It is important to consider the factors of METT-TC when using MP support. During offensive operations, MP assets primarily focus on: planning tasks that support movement of tactical forces throughout the AO; providing support to sustainment forces moving supplies forward; taking control of detainees and DCs to reduce degradation of combat forces.

8-15. In defensive operations, the primary focus for the MP force is to ensure movement of the repositioning or counterattacking forces, provide area security, and support the evacuation of captured or detained individuals. Military police can support stability operations through the five MP functions discussed previously. The primary support they render is law and order. MPs can conduct policing activities in support of civil security and civil control lines of effort. MPs can assist in establishing the rule of law, and internment/resettlement operations focused on controlling and protecting detainees and displaced civilians.

8-16. It is important that MP resources be synchronized and weighted in support of the brigade's main effort. This helps to maximize MP resources allocated to the brigade. MP support might not be available or adequate to perform all necessary MP battlefield functions simultaneously. Commanders must prioritize those missions and designate other elements within the brigade to assist in their execution. When augmented by supporting MP elements, the BCT must understand that not all MP units have similar capabilities. For instance, an MP guard company has approximately 124 personnel; but since this unit is specifically designed to guard facilities, prisoners, or critical assets at static locations, the unit is authorized only nine vehicles. As a result, this unit cannot effectively conduct a mission requiring greater levels of mobility.

## SECTION III – AIR AND MISSILE DEFENSE OPERATIONS

8-17. The BCT does not have organic air defense artillery weapons systems. Air and missile defense (AMD) support to the BCT may be limited. Units apply passive air defense measures and expect to use their organic weapons systems for self-defense against enemy air threats. The BCT commander may request additional air defense assets, if required.

8-18. The brigade does have an organic air defense and airspace management (ADAM)/brigade aviation element (BAE) cell. The ADAM cell is equipped with an AMD workstation (AMDWS), an air defense system integrator, and forward area air defense engagement operations workstation. The BAE is equipped with a tactical airspace integration system (TAIS) workstation.

# ADAM/BAE CAPABILITIES AND FUNCTIONS

8-19. Upon mission receipt, the ADAM cell conducts an assessment to determine if AMD augmentation from higher HQ is required. The cell conducts continuous planning and coordination appropriate for the augmented sensors that the brigade will deploy within its AO. The ADAM cell and tailored AMD augmentation force provide the active air defense across the brigade's distributed force.

8-20. The air defense element of the ADAM/BAE is the commander's expert on organic active and passive air defense operations. The ADAM cell:

- Assists the S-2 in the development of the aerial intelligence preparation of the battlefield.
- Analyzes and makes recommendations on the use of combined arms for air defense and the use of passive air defense measures to protect the force from engagement or observation.
- Participates in fires planning for suppression of enemy air defense (SEAD) and denial of landing/drop zones (LZ/DZ).

8-21. The ADAM/BAE integrates into the joint tactical digital information link network for receiving of and contributing to the aerial common operational picture (COP). The engagement and identification authority for surface to air fires is the chain of control from the ADAM cell to the air defense artillery fire coordination officer at the control and reporting center (CRC) or combined air operations cell.

8-22. The ADAM/BAE communicates the air defense warning and weapons control status (including changes to the local air defense warning) to the BCT, and participates in early warning through electronic means and visual reporting of unknown aircraft. The ADAM/BAE provides early warning and contributes to airspace deconfliction between internal and external ground fires, organic unmanned aircraft systems, (UAS) Army aviation, all other aircraft (military and civilian) and missiles to maximize all airspace users' capabilities, while reducing risk of fratricide and collateral damage.

8-23. The BAE synchronizes Army aviation operations into the BCT scheme of maneuver, integrates aviation into sustainment operations, and represents Army aviation during the MDMP. See FM 3-01.11 for more information on the ADAM/BAE cell.

## COORDINATION AND INTEGRATION OF AIR AND MISSILE DEFENSE AUGMENTATION

8-24. The ADAM cell conducts a supporting METT-TC analysis. Upon completion of this initial analysis, the BCT commander is briefed, and if required, approves the request for air defense augmentation from higher. The BCT may be augmented with an Avenger unit depending on asset availability. A more likely augmentation is the Sentinel radar system, which provides the local air picture and contributes to the aerial COP. Coordination for deployment of the recommended AMD augmentation force runs concurrently with the AMD METT-TC analysis. Depending upon force availability (exclusion area boundary AMD assets already deployed in the AO, the ADAM cell identifies AMD augmentation force requirements and their availability for rapid deployment. It then integrates this information into the AMD force composition recommendation to the BCT commander. Upon approval from the BCT commander, the ADAM cell issues a warning order (WARNO) to the selected AMD augmentation force, which is integrated into the BCT deployment scheme.

8-25. The ADAM cell coordinates between the BCT staff and the AMD augmentation force commander to relate the BCT commander's intent. The cell provides the BCT commander's defended asset list/critical asset list to the AMD augmentation force commander. The cell provides the BCT commander and staff with the aerial component of the overall COP. As the operation evolves, the cell works continuously with the BCT staff to ensure the commander's intent is executed with respect to the aerial COP and defenses. The cell continuously monitors the AMD situation and conducts continual METT-TC analysis to maintain situational awareness of the third dimension in both friendly and enemy perspectives. The ADAM cell merges into the integrated air defense system through cooperation with higher HQ's air defense coordinators (e.g., joint air operations center, combined air operations center, and the battlefield coordination detachment collocated with the United States Air Force (USAF) area air defense commander.

# AIRSPACE MANAGEMENT AND CONTROL

8-26. The ADAM cell receives and distributes the relevant data from the airspace control order (ACO) and air tasking order (ATO), interpreting and displaying the procedural means of airspace control (e.g., corridors, restricted operations zones), and scheduled friendly air operations that can impact BCT operations. In addition, the cell develops recommended airspace control means that support BCT operations and forwards them to the airspace control authority (ACA) for approval and implementation. In all aviation command and control (C2) actions, the ADAM cell coordinates existing and proposed means of airspace control with all elements of the BCT force employing aerial assets (e.g., Army aviation, friendly force UAS, artillery).

8-27. The ADAM/BAE officer in charge (OIC) is the airspace C2 integrator for the S-3. When a BCT is controlling an AO, the authority that the BCT has over Army airspace users is the same as the BCT's authority over ground units transiting its AO. BCTs controlling an AO have authority over all Army airspace users in their AO, as well as joint aircraft in support of BCT operations (such as CAS). All Army airspace users transiting a BCT AO are expected to coordinate with the BCT responsible for the AO they are transiting. Usually, BCTs have the authority to coordinate directly with joint airspace control elements that control airspace over the BCT (CRC/Airborne Warning and Control System) for the purpose of fires coordination or immediate airspace coordination.

## BRIGADE COMBAT TEAM AIRSPACE USERS

8-28. The airspace command and control (AC2) process maximizes the simultaneous use of airspace. At decisive moments, commanders are able to exploit all available combat power—synchronized in time, space, and purpose. The BCT has many users of airspace. The AC2 plan must be integrated, coordinated, deconflicted, and disseminated with each of these users to ensure they do not interfere with each other.

## MOVEMENT AND MANEUVER

8-29. Aviation units can maneuver rapidly to bring aerial firepower, agility, and shock effect at a decisive place and time. Using aviation units to enhance reconnaissance, provide security, and conduct attacks provides the ground force commander with positional advantage over his enemy, and increases the tempo of operations.

8-30. Army special operations aviation units conduct operations throughout the range of military activity. Special operations forces (SOF), due to their mission profile, often operate beyond the usual areas of troop concentrations. Missions deep within enemy territory require the AC2 system to be capable of providing the necessary restrictive operational environment control measures to avoid fratricide.

8-31. Airborne units are subject to many of the same considerations of AC2 as aviation and SOF. While in the air movement phase of the operation, airborne forces require airspace control measures to provide entry and exit routes for the aircraft that deliver forces to their predetermined locations. Airborne operations require restricted operations areas to deconflict airspace from all other aircraft not directly involved in the airborne operation. The ground phase of the operation requires substantial deconfliction of the operational environment.

## INTELLIGENCE

8-32. Maneuver commanders at all levels use aircraft and UAS to gather intelligence. UAS conduct intelligence-collection and target-acquisition missions over the entire battlefield. Missions for UAS at higher altitudes are included in the ATO/ACO. BCT UAS can be covered by the ATO/ACO. The BCT ADAM/BAE cell must ensure that continuous coordination for the employment of BCT UAS is conducted with higher headquarters. These missions are planned for inclusion in the ATO and the ACO. However, because of their flexible, highly responsive nature, UAS assets are often tasked for immediate missions that are not in the ATO or ACO. The ADAM/BAE cell at each BCT must resolve conflicts between UAS assets and those of other airspace users. The S-2 provides the information required for coordinating intelligence collection missions with the ADAM/BAE cell to synchronize these missions with other airspace operational requirements, especially air defense forces.

## FIRES

8-33. Field artillery uses airspace to deliver indirect fire support to maneuver forces across the entire area of the distributed battlefield. These indirect fires can traverse the airspace from extremely low to very high altitudes. All planned artillery fires are coordinated with other airspace users. However, not all targets can be identified and fires deconflicted in advance. In the close battle, fires of an unplanned, immediate nature in response to the actions of the maneuver forces, and the reaction by the enemy occur. Commanders should indentify the potential of fratricide in these instances and incorporate guidance to minimize and control risk by implementing preventive measures.

## SUSTAINMENT

8-34. Aero-medical evacuation and/or casualty evacuation (CASEVAC) provides speed, range, and flexibility to move patients directly to a medical treatment facility that is best equipped to treat the casualty. Though the aviation brigade is responsible for command, control, and execution of the aero-medical evacuation mission, the BCT ADAM/BAE cell should be ready to synchronize this mission with other airspace users within the BCT AO. Refer to FM 4-02.2 for additional information on casualty evacuation and medical evacuation.

## PROTECTION

8-35. AMD units may be located throughout the combat zone area of operations to defend maneuver forces, vital airfields, logistics elements, and other critical assets as prioritized by the commander. AMD forces use both positive and procedural means of fire control for air battle C2. Close integration between airspace control, other airspace users, and air defense C2 is imperative to ensure safe, unencumbered passage of friendly aircraft while denying access to enemy aircraft and missiles.

## OTHER AIRSPACE USERS

8-36. Other users of BCT airspace may include: Army tactical missile systems, close air support, host nation aircraft, counter-fire artillery, routine aerial resupply missions, and the enemy's fixed- and rotary-wing aircraft, UAS, theater ballistic missiles, and missiles.

## SECTION IV – CHEMICAL, BIOLOGICAL, RADIOLOGICAL, AND NUCLEAR OPERATIONS

# CBRN STAFF SECTION

8-37. The brigade CBRN staff section advises the commander on all CBRN matters. The chemical CBRN section is responsible for collecting, consolidating, and distributing all CBRN reports from subordinate, adjacent, and higher units. The chemical CBRN section inspects chemical CBRN equipment and trains subordinate units on CBRN defensive tasks prior to deployment and in garrison. The BCT CBRN officer is a member of the operations staff officer (S-3) plans and operations section and usually operates in the main command post (CP).

8-38. The BCT CBRN officer acts as the liaison with any attached CBRN elements. He is required to coordinate closely with the S-2 on the current and updated CBRN threat. Together they develop CBRN named areas of interest (NAI). The CBRN officer coordinates with fire support and aviation personnel on planned obscuration operations and advises them of hazard areas. He also coordinates with the logistics staff officer (S-4) on CBRN logistics matters (e.g., mission-oriented protective posture [MOPP], protective mask filters, fog oil), and to identify both "clean" and "dirty" routes as well as contaminated casualty collection points. The CBRN officer exercises staff supervision over the CBRN reconnaissance platoon in the BSTB and synchronizes their activities with reconnaissance planning.

# CBRN RECONNAISSANCE

8-39. The CBRN reconnaissance platoon informs the commander of chemical, biological, radiological, and nuclear obstacles on the battlefield. The purpose of CBRN reconnaissance is to detect, identify, report, and mark specific CBRN hazards. Additional information on CBRN reconnaissance can be found in FM 3-11.19.

## CBRN RECONNAISSANCE IN THE OFFENSE

8-40. In the offense, maneuver elements must be able to maintain agility and get to the right place at the right time. Enemy forces may use CBRN weapons to slow down or impede attacking friendly forces. Use of CBRN weapons can disrupt the tempo and momentum of the attack, allowing the enemy to regain the initiative. Our forces employ CBRN reconnaissance to maintain the freedom of maneuver for combat forces on axes of advance, main supply routes (MSR), and critical areas that the commander identified. Since CBRN reconnaissance focuses on intelligence preparation of the battlefield (IPB) , it is integrated into the intelligence, surveillance, and reconnaissance (ISR) plan. CBRN reconnaissance personnel are responsible for covering specific NAIs for specific periods of time. In the offense, CBRN reconnaissance is focused on operations that provide the commander with flexibility, retain freedom of maneuver, and identify known or suspected areas of contamination.

## CBRN RECONNAISSANCE IN THE DEFENSE

8-41. In the defense, as in the offense, CBRN reconnaissance focuses on IPB. In defensive operations, however, CBRN assets should focus on ensuring freedom of movement along brigade routes of reinforcement, forward and rearward mobility corridors, and other critical areas identified by the commander. CBRN reconnaissance elements in the defense can conduct route reconnaissance, confirm or deny suspected or known CBRN hazards at NAIs, perform reconnaissance as part of quartering party operations, support counterattacking forces, and conduct CBRN surveillance of battle positions with standoff detection.

## SECTION V – MILITARY INTELLIGENCE COMPANY SUPPORT

8-42. The MICO mission is to conduct analysis, intelligence synchronization, full motion video, signals intelligence (SIGINT and human intelligence (HUMINT) collection. MICO does this in support of the BCT and its subordinate commands across the full spectrum of operations. The MICO provides analysis and intelligence synchronization support to the BCT S-2. The MICO supports the BCT and its subordinate commands through collection, analysis, and dissemination of intelligence information. It supports the BCT S-2 in ISR synchronization and in maintaining a timely and accurate picture of the enemy situation. See FM 2-19.4 for more detail on the activities of the MICO.

# EMPLOYMENT AND PLANNING CONSIDERATIONS

8-43. During the brigade's planning, the MICO commander, acting as the BCT intelligence collection manager, assists the brigade S-2 with the development of the intelligence running estimate and all intelligence products and deliverables needed to support the brigade orders process. These include but are not limited to the mission analysis briefing, base operation order (OPORD) input, and annex B. When the BCT commander approved the brigade order, the MICO commander produces the company OPORD In addition to task organization considerations in FM 5-0, the MICO commander:

- Provides seamless analytical support to the brigade S-2.
- Assists with the synchronization of intelligence and electronic warfare (IEW) assets in the brigade's AO.
- Reallocates and repositions company assets in response to changes in the brigade's mission, concept of operations, scheme of support, and threat.
- Establishes logistics and security relationships with the brigade headquarters and headquarters company (HHC) to sustain and protect the MICO personnel and equipment.
- Integrates other attachments into company operations as directed in the brigade order.

8-44. The ISR requirements section and the situation and target development section, of the analysis and integration platoon usually operate under the BCT S-2's operational control (OPCON). The tactical unmanned aircraft system (TUAS) platoon and the ground collection platoon assets may be deployed within the BCT's AO under differing command and support relationships. The SBCT's reconnaissance squadron assets may deploy under differing command and support relationships that may also require similar coordination and planning. These relationships may require the MICO commander to conduct logistical and security coordination and planning with other brigade C2 elements.

8-45. Command relationships establish the degree of control and responsibility a commander has for the forces operating under his or her control. Command relationships can be attached, under OPCON or tactical control (TACON). HUMINT collection teams from the ground collection platoon may operate in direct support of subordinate brigade elements in combat operations.

# PERSONNEL AND EQUIPMENT

8-46. The organization of the MICO in the HBCT and IBCT is somewhat different than the organization of the SBCT MICO. These companies contain a headquarters element, an analysis and integration platoon, a TUAS platoon, and a ground collection platoon. They conduct ISR integration and intelligence production in support of the brigade's planning, preparation, and execution of multiple, simultaneous decisive actions across the distributed AO (Figure 8-2).

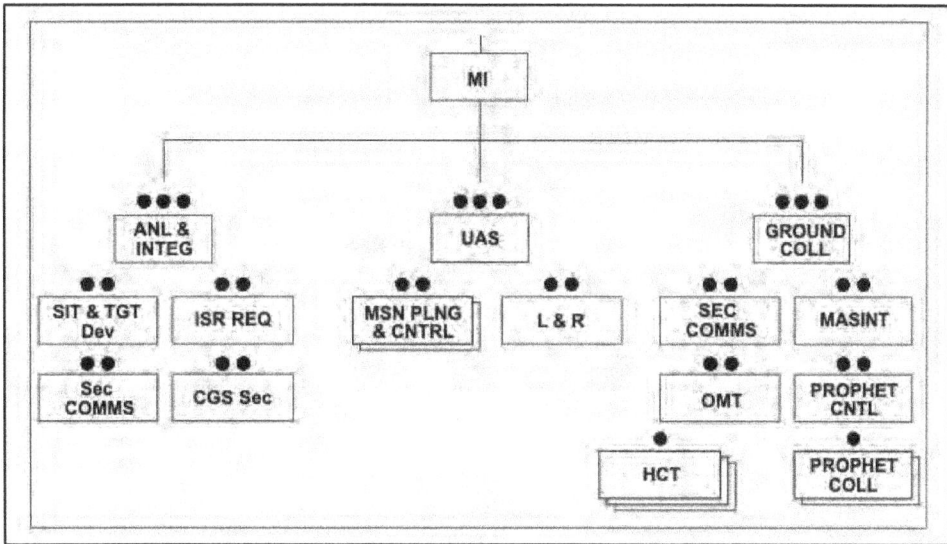

**Figure 8-2. Heavy and Infantry BCT military intelligence company**

8-47. The SBCT MICO has an analysis platoon and an ISR integration platoon. It does not have a tactical UAS platoon. The SBCT TUAS capability is found in the surveillance platoon of the reconnaissance squadron.

## COMPANY HEADQUARTERS

8-48. The MICO commander directs the employment of the company in accordance with assigned missions and guidance from the brigade HQ. The MICO commander must position himself where he can fulfill all of his command responsibilities best. This position could be in the brigade CP, or on site with a HUMINT collection team. The commander's location could also be at the reconnaissance squadron's CP or with a supported maneuver commander at a critical time and location on the battlefield. From any of these locations, the MICO commander will conduct battlefield circulation, maintain situational awareness of all teams' positions, and perform required administrative functions.

8-49. The MICO CP usually is collocated with the brigade main tactical operations center (TOC) to facilitate C2 of the company assets and to maximize BCT S-2 support. The MICO CP includes the company headquarters element, the analysis and integration platoon, the TUAS platoon, and the ground collection platoon. During brigade operations, the situation and target development section and ISR requirements section are typically under OPCON of the BCT. Human intelligence collection teams of the ground collection platoon may operate in direct support of a maneuver battalion, the reconnaissance squadron, or even to their subordinate companies or troops.

## PLATOON HEADQUARTERS

8-50. The platoon headquarters usually is located where they can best C2 their platoon elements. All three platoons—analysis and integration platoon, TUAS platoon, and ground collection platoon—usually are collocated with the BCT S-2. Elements of the TUAS platoon and ground collection platoon could be deployed anywhere in the AO.

## ANALYSIS AND INTEGRATION PLATOON

8-51. The analysis and integration platoon provides the BCT S-2 analytical support. It consists of a headquarters section, a situation and target development section, an ISR requirements section, a common ground station section, and a satellite communications (SATCOM) )team. The ISR requirements section and the situation and target development section collocate with the brigade main CP, and are OPCON to the BCT S-2. They provide the BCT S-2 automated intelligence processing, analysis, and dissemination capabilities as well as access to intelligence products of higher echelons.

## TACTICAL UNMANNED AIRCRAFT SYSTEM PLATOON

8-52. The MICO in the HBCT and the IBCT, each have a TUAS platoon. In the SBCT, the surveillance troop of the reconnaissance squadron has a TUAS platoon. The TUAS platoon consists of one mission planning and control section, and one launch and recovery section. The platoon is equipped with four RQ-7 Shadow aircraft. The Shadow UAS enhances tactical level reconnaissance, surveillance, target acquisition, and the commander's battle damage assessment (BDA). All UAS missions are synchronized with the air defense airspace management cell.

8-53. The RQ-7 Shadow is a small, lightweight TUAS. The system is comprised of air vehicles, modular mission payloads, ground control stations, launch and recovery equipment, and communications equipment. It carries enough supplies and spares for an initial 72 hours of operation. It is transportable in two high-mobility multipurpose wheeled vehicles (HMMWV) with shelters, and two additional HMMWVs with trailers as troop carriers.

8-54. The platoon includes three Shadow aircraft with a fourth aircraft as part of the issued equipment of the maintenance section. The payload has a commercially available electro-optic and infrared camera, a communications relay package module, and communications equipment for C2 and imagery dissemination. Onboard global positioning system instrumentation provides navigation information.

## GROUND COLLECTION PLATOON

8-55. The ground collection platoon contains a tactical HUMINT section and a Prophet control section. The HUMINT section collects HUMINT through screening interrogations, debriefing contact operations, and support to document and media exploitation activities. The HUMINT section coordinates and executes HUMINT operations as directed by the brigade S-3 in coordination with the brigade S-2 and S-2X. The Prophet control section coordinates and executes signals intelligence operations as directed by the brigade S-3 in coordination with the brigade S-2. The HUMINT section contains human intelligence collection teams that conduct source operations, write reports, and collect in support of their respective maneuver unit's intent and priority intelligence requirements (PIR). The human intelligence collection teams are usually direct support to maneuver units in the area of operation. Each human intelligence collection team is comprised of four team members and two interpreters. The operational management team manages the teams, deconflicts sources, reports to the S-2X and justifies the use of the HCTs in the AO.

## TROJAN SPIRIT LITE

8-56. The SATCOM team is responsible for integrating sensitive, compartmented information (SCI) communications into the existing network architecture using the TROJAN SPIRIT lite system. The TROJAN SPIRIT lite, organic to the analysis and integration platoon, usually collocates with the BCT S-2 and the analysis and integration platoon during operations. The intelligence analysts assigned to the TROJAN SPIRIT lite access the dedicated multilevel security, high-capacity communications link between BCT CPs, theater, joint centers, national centers, and other intelligence organizations outside the BCT's AO.

They do this to pull intelligence products, receive and analyze routed direct downlinks, and access external databases to fuse with organically collected information. The system also provides the opportunity for secure, analytic collaboration externally to the BCT. The TROJAN SPIRIT lite also provides access to the joint worldwide intelligence communications system through its joint deployable intelligence support system.

## BRIGADE WEATHER TEAM (USAF)

8-57. The brigade weather team of the USAF, when attached, provides the BCT with a weather prediction and weather effects analysis capability. The brigade weather team:

- Evaluates and applies operational weather squadron forecasts to specific brigade missions, weapons systems, strategies, tactics, and applications; deploys with the brigade; and in general provides both direct and indirect tailored customer support.
- Provides meteorological support to brigade planning, training, deployment, employment, and evaluation.

8-58. In addition to support to the S-2, the brigade weather team is the main source of weather support for all brigade warfighting functions. As a member of the commander's special staff, the staff weather officer is responsible for coordinating operational weather squadron and service matters through the S-2. The staff weather officer is the weather liaison between Army customers and the Air Force forecasting resources developed at centralized (regional) production centers.

## SECTION VI – INFORMATION ENGAGEMENT

8-59. Information engagement is the integrated employment of public affairs (PA), MISO, combat cameras, and leader and Soldier engagements to support both public information and diplomacy. This section describes how the BCT and information-related augmentation units (e.g., tactical MISO teams) accomplish information engagement. The S-7 coordinates information engagement with a staff organization within the BCT dedicated to this role. The rest of the BCT staff coordinate other elements of joint information operations such as:

- Fires and command, control, communications, and computer operations staff elements coordinate command and control warfare.
- The command, control, communications, and computer operations section and the signal staff officer (S-6) manage information protection.
- Operations security (OPSEC) and military deception is an operations function (S-3) with intelligence support (S-2).

# ORGANIZATION

8-60. Each BCT staff has a group of staff sections that contribute to information engagement (IE) activities across the brigade. The lead staff section is the BCT S-7 who manages the information operations (IO) cell. Other BCT elements that may contribute and collaborate on IE tasks are:

- Public affairs.
- MISO.
- Civil affairs (CA) operations staff sections and teams.
- Combat camera.
- HUMINT collection teams.
- Special operations liaison.
- Criminal intelligence division liaison.
- Host nation liaison and/or interpreter.
- Contractor subject matter experts.

8-61. When so directed, these BCT elements form a working group to coordinate nonlethal operations. For stability operations, information engagement staff may be supplemented by other officers out of the fires, intelligence and/or sustainment staff elements. A smaller, but parallel staff organization for information engagement exists in most of the subordinate battalions of the BCT. In counterinsurgency operations, company and even platoon HQ may organize a small element dedicated to coordinating information engagement for their AOs. All echelons participate in collaborative, parallel, and multi-echelon planning sessions within the information engagement working group. Staff officers coordinate information engagement; leaders and Soldiers execute information engagement tasks.

## ARMY INFORMATION TASKS

8-62. FM 3-0 directs how the Army conducts activities called IO at the joint level. The Army conducts Army information tasks that reorganize joint IO responsibilities into:

- Information engagement.
- C2 warfare.
- Information protection.
- OPSEC.
- Military deception.

# PLANNING

## ELEMENTS OF INFORMATION ENGAGEMENT

8-63. Information engagement supports the commander in achieving his end state through coordinating the following:

- Inform and educate internal and external publics.
- Influence the behavior of target audiences.

8-64. Information engagement uses the following capabilities:

- Leader engagement.
- Soldier engagement.
- Public affairs.
- MISO.
- COMCAM.
- Strategic communications and defense support to public diplomacy.

## LEADER AND SOLDIER ENGAGEMENT

8-65. Face-to-face interaction by leaders and Soldiers strongly influences the perceptions of the local populace. Carried out with discipline and professionalism, day-to-day interaction of Soldiers with the local populace among whom they operate has positive effects. Such interaction amplifies positive actions, counters enemy propaganda, and increases goodwill and support for the friendly mission. Likewise, meetings conducted by leaders with key communicators, civilian leaders, or others whose perceptions, decisions, and actions will affect mission accomplishment can be critical to mission success. These meetings provide the most convincing venue for conveying positive information, assuaging fears, inducing cooperation, and refuting rumors, lies, and misinformation. Conducted with planning and preparation, both activities often prove crucial in assessing the local situation, garnering local support for Army operations, providing an opportunity for persuasion, and reducing friction and mistrust.

8-66. Soldier engagement includes tactical tasks such as:

- Project a positive and stabilizing image to the population while on patrols or manning checkpoints.
- Assess the mood on the street among the population in the patrol report.
- Pass written or oral messages to neighborhoods and businesses.
- Collect information requirements related to information engagement.
- Administer surveys to the local population.

8-67. Leader engagement includes tactical tasks such as:

- Talk with key contacts to discover local requirements and the power structures in place among local society.
- Establish points of contact between the leaders of each echelon of BCT forces and appropriate levels of authority within the local leader network.
- Negotiate with local leaders to gather information, pass on themes and messages, assess the environment, and accomplish information engagement effects.
- Manage relationships between the military and the diverse group of local leaders and the media to accomplish the commander's end state with respect to information engagement and popular support.
- Be the face of the U.S. military in the BCT AO (e.g., give press conferences, attend public presentations, and accompany presence patrols).

## PUBLIC AFFAIRS

8-68. PA is a commander's responsibility to execute public information, command information, and community engagement directed toward both the external and internal publics with interest in BCT operations. PA proactively informs and educates internal and external publics through public information, command information, and direct community engagement. Although all information engagement activities are completely truthful, PA is unique. It has a statutory responsibility to factually and accurately inform various publics without intent to propagandize or manipulate public opinion. Specifically, PA facilitates the commander's obligation to support informed U.S. citizenry, U.S. Government decision makers, and non-U.S. audiences. Effective information engagement requires particular attention to clearly demarking this unique role of PA by protecting its credibility. This requires care and consideration when synchronizing PA with other information engagement activities. PA and other information engagement tasks are synchronized to ensure consistency, command credibility, and OPSEC. The PA staff performs the following:

- Advise and counsel the commander concerning PA.
- Develop PA execution information.
- Perform media facilitation.
- Train the BCT on PA topics.
- Coordinate community engagement.
- Develop BCT communications strategies.

## MILITARY INFORMATION SUPPORT OPERATIONS

8-69. MISO are planned operations that convey selected information and indicators to foreign audiences. The purpose of MISO is to influence the audiences' emotions, motives, objective reasoning, and ultimately the behavior of foreign local governments, organizations, groups, and individuals in the BCT AO. The MISO goal is to induce or reinforce local attitudes and behavior favorable to the BCT commander's objectives. Commanders focus MISO efforts on adversaries, their supporters, and their potential supporters. They may integrate these capabilities into the operations process through information engagement and the targeting process. Tactical MISO teams may be task-organized with BCT elements to perform information engagement tasks.

## COMBAT CAMERA

8-70. COMCAM is the acquisition and utilization of still and motion imagery in support of operations involving the BCT. Such operations include combat, information, humanitarian, intelligence, reconnaissance, engineering, legal, PA, and others. COMCAM teams generate still and video imagery in support of BCT intelligence collection and information engagement activity. For example, COMCAM teams can prepare products documenting Army tactical successes that counter enemy propaganda claiming the opposite. It can provide targeted footage to media of BCT actions not observed by the civilian press.

STRATEGIC COMMUNICATIONS AND DEFENSE SUPPORT TO PUBLIC DIPLOMACY

8-71. Strategic communications involves focused U.S. Government efforts to understand and engage key audiences to create, strengthen, or preserve conditions favorable for the advancement of U.S. Government interests, policies, and objectives. This is accomplished through the use of coordinated programs, plans, themes, messages, and products synchronized with the actions of all instruments of national power (Joint Publication [JP] 1-02). While U.S. leaders communicate some of this information directly through policy and directives, they also shape the environment by providing access and information to the media. The BCT commander supports strategic communications as required within his AO normally through public affairs and leader engagement tasks.

8-72. Defense support to public diplomacy are those activities and measures taken by the BCT to support and facilitate public diplomacy efforts of the U.S. Government representative. Defense support to public diplomacy is a key military role in supporting the U.S. Government's strategic communications program. It includes peacetime military engagement activities conducted as part of combatant commanders' theater security cooperation plans. These higher headquarters requirements are often addressed using leader and Soldier engagement, MISO, and PA tasks.

## SECTION VII – CIVIL AFFAIRS ACTIVITIES

8-73. CA activities establish and maintain relations among U.S. military forces, host nation, nongovernmental organizations (NGO), U.S. state department agencies, other U.S. Governmental agencies, and the civilian population. Favorable relations among these participants facilitate BCT operations. CA activities play a supporting role to information engagement and are a key element in stability operations. Effective CA activity requires close cooperation with national, international, and local interagency partners. These partners are not under military control. Many NGOs, for example, do not want to be too closely associated with military forces because they need to preserve their perceived neutrality. Interagency cooperation may involve a shared analysis of the problem building a consensus that allows synchronization of military and interagency efforts. The military's role is to provide protection, identify needs, facilitate CA activity, and use improvements in social conditions as leverage to build networks and mobilize the populace.

# CIVIL AFFAIRS INTEGRATION INTO THE BRIGADE COMBAT TEAM

8-74. The BCT CA staff officer (S-9) is responsible for CA activity planning. The S-9 usually conducts collaborative planning as part of the information engagement working group or targeting working group. The S-9 has a close working relationship with sustainment (including contracting) and engineer units within the BCT for the conduct of CA projects. CA projects are usually tasks assigned to subordinate commanders owning the AO. The BCT CA staff helps coordinate planning and support for these projects. The BCT civil-military operations (CMO) also organizes the BCT civil-military operations center (CMOC).

8-75. Depending on the factors of METT-TC, the CA company HQ may participate in the BCT's planning processes if directed. Usually, the HQ is attached to the brigade for planning, operations, security, and sustainment. CA teams often are attached to subordinate battalions to become their CA maneuver element and their CMO staff element. Their role can include organizing battalion CMOCs.

# CIVIL AFFAIRS UNITS

8-76. CA units are those units organized, trained and equipped to conduct CA operations. CA units provide the commander with the means to shape his operational environment (OE) with regard to these significant factors and to synchronize their actions with those of the military force. In addition, CA units perform important liaison functions between the military force and the local civil authorities, international organizations, and NGOs. A CA company may be attached to support a BCT (Figure 8-3).

- CA personnel engage in a variety of CA operations in fulfillment of CA core tasks. CA elements can assess the needs of civil authorities, act as an interface between civil authorities and the military supporting agency, and as liaison to the civil populace. They can develop population and resource control measures and coordinate with international support agencies.

- CA unit personnel are regionally oriented and possess cultural and linguistic knowledge of the countries in each region. With guidance from the commander on desired effects, CA personnel have a wide variety of resources at their disposal to influence the AO. CA is a combat multiplier in this sense. In addition, the civilian skills that reserve component CA units possess enable them to assess and coordinate infrastructure activities. See FM 3-05.401 for more details.

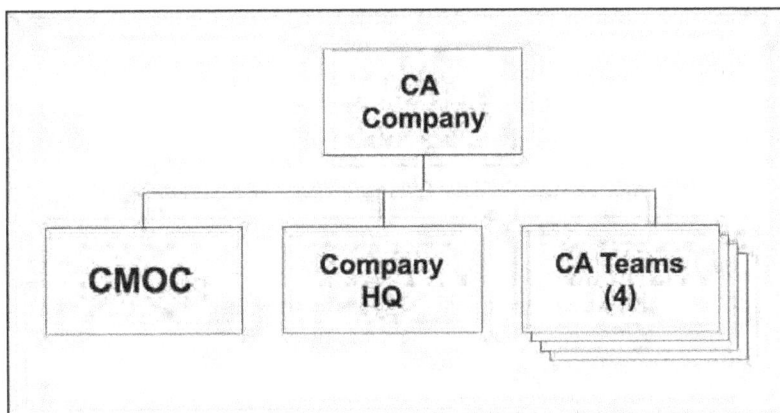

**Figure 8-3. Civil affairs company**

## CIVIL-MILITARY OPERATIONS CENTER

8-77. A CMOC is a coordination center that any commander establishes when required. CMOCs assist in coordinating the activities of the commander's military forces, other U.S. Government agencies, NGOs, and other international organizations. There is an established CMOC structure, but its size and composition are situation dependent (Figure 8-4). The CMOC serves as the primary center for synchronizing military operations with the operations of nonmilitary organizations. During transition from military to civilian control, the CMOC can serve as a source of operational continuity and a facilitator to the transition process.

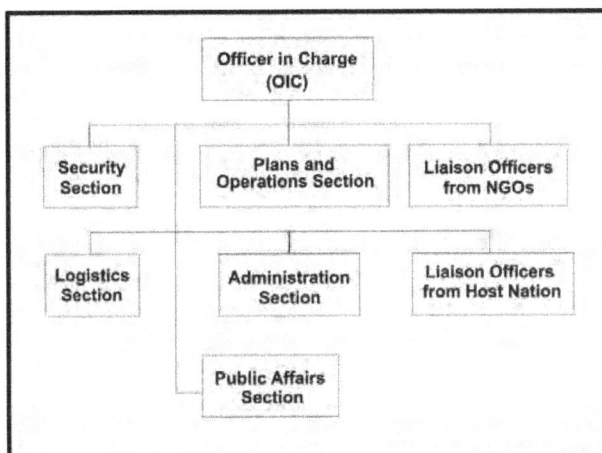

**Figure 8-4. Basic civil-military operations center structure**

8-78. A CMOC has an OIC or a noncommissioned officer in charge (NCOIC) who is responsible for overall management of CMOC operations. The BCT commander may assign his S-9 as the OIC if there is no CA company in support. Depending on METT-TC, the OIC/NCOIC might be from an attached CA unit, a senior civilian, or a foreign military officer.

8-79. The CMOC typically has both civilian-related and military-related staff sections:

- The public affairs branch handles media inquiries to coordinate the release of information to the public with the information engagement staff officer (S-7), and to synchronize CMOC information with the BCT's S-9 section.
- The security branch manages the various aspects of security (physical, operations, personnel, and information) inherent to CMOC operations.
- Liaison officers or representatives are on-site CMOC contacts for both military and civilian agencies and/organizations.
- The plans and operations section maintains the current status of projects and routes used by CA activities.
- The logistics section maintains a database of all points of contact and host nation resources that can be used for military or humanitarian purposes (e.g., facilities, transportation assets, goods, services). Generally, this section tracks costs incurred by military forces and other agencies participating in CA activities. It often includes a contracting officer element.

8-80. The administration section focuses on internal CMOC activities and personnel issues that include:

- Maintaining access rosters and identification systems for the CMOC.
- Conducting CMOC meetings, minutes, and scheduling.
- Processing and archiving required reports.
- Creating recognition documents and certificates.
- Maintaining adequate levels of supplies for use in CMOC operations (e.g., office supplies, fuel, batteries, and light bulbs).
- Managing operator-level maintenance on vehicles, communications, and generator equipment.

8-81. The establishment of a CMOC must be a carefully considered balance between security and utility. Facilities and information systems (INFOSYS) must be procured and sustained, and security provided for personnel, equipment, and information. It can be located at a convenient local government facility centrally located to CA requirements or on the edge of a friendly forces perimeter where it can be secured but is still accessible to non-military personnel. The BCT HQ usually establishes at least one CMOC, and often each battalion will establish a CMOC as well when it owns part of the AO. In some cases, company HQ may establish a small CMOC to help conduct CA in their AO.

## SECTION VIII – COMMAND AND CONTROL INFORMATION SYSTEMS

8-82. C2 information systems are more critical now than in the past due to increasing bandwidth requirements at increasingly lower levels of the command. The BCT commander must consider the reliability of communications in determining the level of acceptable risk when allocating network operations elements during an operation. The correct placement of limited communications resources provides the BCT with the capability to transfer great volumes of information at high rates of speed to all the required recipients.

# BRIGADE SIGNAL COMPANY

8-83. The brigade signal company is organic to the BCT and provides the primary communications support to the brigade. The brigade signal company deploys, installs, operates, and maintains the C2 networks that support brigade operations and integrate with external networks. The headquarters and network support platoon and the network extension platoons are the operational arms of the company. They deploy and operate the brigade transmission and switching systems to provide voice, data, and network installation and management support:

8-84. The brigade signal company:

- Provides reach back connectivity through the division HQ.
- Provides range extension of the brigade's communications services.
- Provides network management.
- Establishes CP voice and/or video capabilities.
- Performs limited signal electronic maintenance.

8-85. Each brigade signal company has two network extension platoons. Each platoon consists of a joint network node section and an extension section. These elements enable the BCT with line of sight and beyond line of sight connectivity to provide unclassified, secret, and top secret/sensitive compartmentalized information (TS/SCI) voice and data, tactical network coverage, and command post support in the BCT AO. Usually one network extension platoon is located at the BCT main CP and one at the brigade support battalion (BSB) CP.

8-86. The brigade signal company, company HQ consists of the command section, the wireless network extension team, and a small CP support team.

- Command section. The command section consists of the company commander, first sergeant, and supply NCO. The command section is responsible for the administration and logistics support for the company.
- Wireless network extension team. The wireless network extension team provides beyond line of sight connectivity for the various CPs and command and control nodes.
- CP support team. Generally the CP support team deploys with the BCT tactical command post (TAC CP) to provide communications connectivity that is similar to, but on a lesser scale than, the network extension that platoons provide to the main and BSB CPs.

# NETWORK SYSTEMS PLANNING

8-87. The S-6 must conduct an electronic preparation of the battlefield (EPB) early enough for the commander to make informed decisions about assigning missions to reconnaissance and surveillance assets and, subsequently, to maneuver forces. A risk analysis based on the recommended network architecture is critical to the EPB.

8-88. The commander must incorporate the EPB into his decision-making process early enough to understand the limitations in communications when planning the maneuver for his unit. The commander must also indicate what his critical information requirements are in order for the S-6 to ensure infrastructure support to that requirement. The S-6 refines the initial EPB as the commander decides what risks he will accept in command, control, communications, and computer operations. The S-6 plans the area of operations' coverage with the available networks. The S-6 identifies any shortfalls in communications coverage, and notifies the S-3 and commander. The commander's estimate of critical information requirements determines the refinement of coverage.

8-89. Adapting the network systems plan to the priorities set by the commander requires close coordination by the S-6 with other staff members and particularly with the executive officer (XO). The XO determines the priority of information being passed, and the S-6 provides the transport path for that information. The XO must ensure the S-6 is aware of the information priorities at all times, and the S-6 must ensure the XO is aware of the system limitations and capabilities. The BCT commander expresses his standard operating procedures for information management in the command information management plan, which the S-6 maintains.

8-90. The signal annex must provide clear understanding of the unit's communications architecture, and how it will operate on the battlefield. A number of information presentation styles are effective: text, preformatted templates, and matrixes. The annex must incorporate all communications resources. The way to do this best is to provide the commander with a signal concept sketch. The graphic presentation provides the commander with a clear and concise understanding of the communications plan. Critical elements include concept of communications, command post locations, and tactical range extension locations. The S-6 must capture information for the complete task organization to portray an accurate picture.

## SECTION IX – EXTERNAL AUGMENTATION TO THE BRIGADE COMBAT TEAM

## SUPPORT AND FUNCTIONAL BRIGADES

8-91. A mix of other brigade types supports the BCT and carries out specific tasks in support of echelons above BCT. The five supporting brigade types are the battlefield surveillance brigade (BFSB), fires brigade, combat aviation brigade, MEB, and sustainment brigade (Figure 8-5).

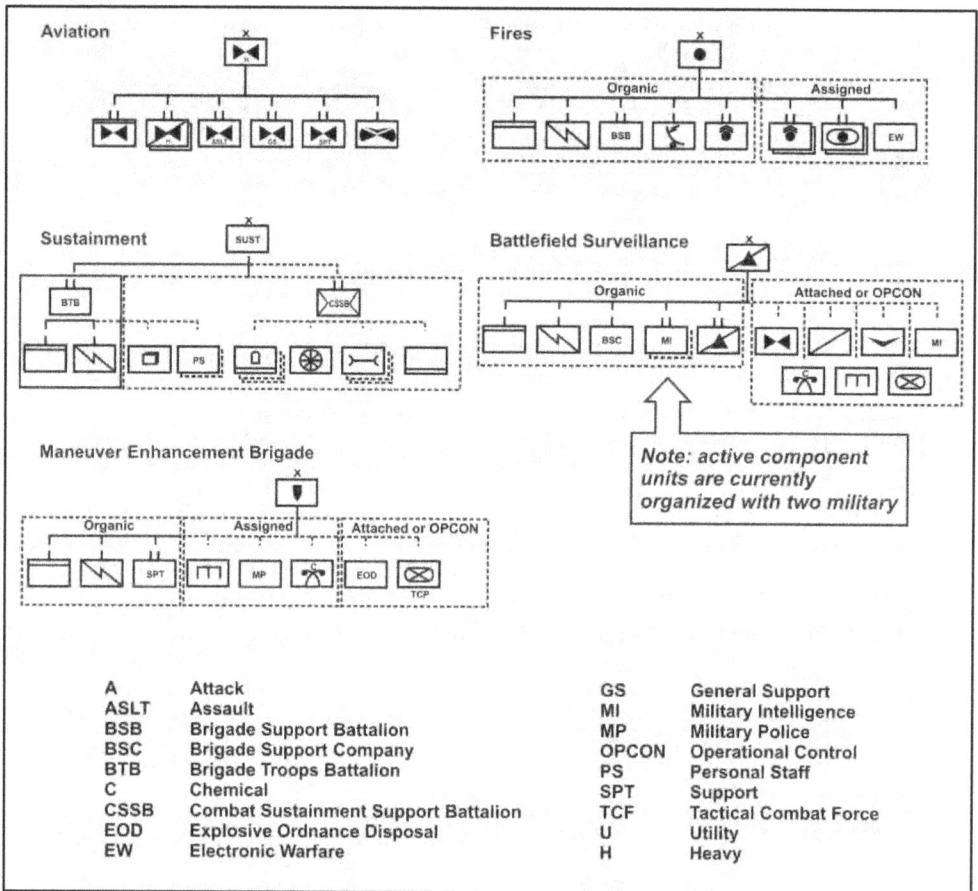

Note: active component units are currently organized with two military

| A | Attack | GS | General Support |
|---|---|---|---|
| ASLT | Assault | MI | Military Intelligence |
| BSB | Brigade Support Battalion | MP | Military Police |
| BSC | Brigade Support Company | OPCON | Operational Control |
| BTB | Brigade Troops Battalion | PS | Personal Staff |
| C | Chemical | SPT | Support |
| CSSB | Combat Sustainment Support Battalion | TCF | Tactical Combat Force |
| EOD | Explosive Ordnance Disposal | U | Utility |
| EW | Electronic Warfare | H | Heavy |

Figure 8-5. Support brigades

### BATTLEFIELD SURVEILLANCE BRIGADE

8-92. The BFSB primarily conducts reconnaissance and surveillance (including MI discipline collection) tasks. It can be assigned to support a division, corps, joint task force (JTF), other Service, or a multinational force. The information it collects focuses on the enemy, terrain and weather, and civil considerations factors of METT-TC to feed the COP. BFSB collection efforts focus on unassigned areas in a higher or supported HQ's AO. The supported commander assigns the BFSB an AO. The BFSB commander does not control BCT reconnaissance and surveillance assets. Each BCT retains control of its organic collection assets and collects information in its assigned AO.

## FIRES BRIGADE

8-93. The fires brigade performs tasks previously executed by division artillery, field artillery brigades, and corps artillery. The fires brigade conducts combined arms operations to provide fires in support of the commander's operational and tactical objectives. The fires brigade executes most Army and joint fires in unassigned areas within the division AO. It also provides reinforcing fires in support of BCT operations. It can use Army, and joint surface- and air-delivered fires as well as incorporating SOF, electronic warfare, and AC2 elements. The fires brigade gives the supported commander a HQ to conduct strike, counterfire, and reinforcing fires throughout the supported HQ's AO.

8-94. Fires brigades have the ability to reconnoiter, detect, and attack targets, and to confirm the effectiveness of their fires. They have networked intelligence, robust communications, and systems that facilitate effective fires. The fires brigade can be a supported or supporting unit and provide and coordinate joint lethal and nonlethal fires including electronic warfare. Fires brigades also have the necessary fires and targeting structure to effectively execute the entire decide, detect, deliver, and assess process. The fires brigade provides:

- A field artillery HQ for the full complement of Army and joint lethal and nonlethal fires (if augmented by AF tactical control party).
- Close reinforcing fires, counterfire, UAS, and counterbattery radar coverage.

## COMBAT AVIATION BRIGADE

8-95. The combat aviation brigade is the primary aviation force provider within the division. The role of the aviation brigade is to support ground maneuver through aviation operations. The brigade can support the BCT's use of pure or task-organized units, and conduct multiple independent missions. Aviation brigade missions include:

- Reconnaissance.
- Security.
- Movement to contact.
- Attack.
- Air assault.
- Air movement.
- C2 support.
- Aeromedical evacuation.
- Aerial CASEVAC.
- Personnel recovery.

8-96. The aviation brigade's attack reconnaissance aircraft deploys to support the BCT commander's scheme of maneuver, and to extend the BCT's operational environment significantly. Attack reconnaissance aircraft can:

- Assist in locating the threat.
- Building and sharing the COP.
- Enhancing force protection.
- Enabling freedom of movement.
- Facilitating unobstructed movement for air assault (AASLT) missions.
- Securing routes for aerial/ground resupply.
- Allowing the commander to focus combat power at the decisive point and time.

8-97. Sensor video recording capability can provide the BCT commander with excellent reconnaissance and BDA information. FM 3-04.126 provides detailed information about attack reconnaissance operations and planning.

8-98. The aviation brigade's utility and heavy helicopter assets enable the maneuver commander to sustain continuous offensive or defensive operations, and to conduct brigade level AASLTs. AASLT operations extend the tactical reach of the maneuver commander, negate effects of terrain, seize key nodes, gain the advantage of surprise, and dislocate or isolate the enemy. Forward air reserve points (FARP) emplaced by lift aircraft and ground assets enable aviation to support and sustain operations throughout the AO. Additionally, heavy lift helicopters are capable of transporting internal and external cargo in a variety of configurations to meet the BCT's sustainment requirements. FM 3-04.113 and FM 90-4 provide detailed information about AASLT operations, and utility and heavy helicopter planning.

8-99. Though not part of the aviation brigade, the BCT's BAE is essential to incorporating aviation into the ground commander's scheme of maneuver. The BAE focuses on providing employment advice and initial planning for aviation missions, UAS, airspace planning and coordination, and synchronization with the air liaison officer and the BCT fire support officer. The BAE also coordinates directly with the aviation brigade or the supporting aviation task force for detailed mission planning. Additionally, it advises the BCT commander and staff on the status and availability of aviation assets, and their capabilities and limitations. Refer to TC 1-400 for additional information about the roles of the BAE.

## MANEUVER ENHANCEMENT BRIGADE

8-100. The MEB is a unique multifunctional C2 headquarters designed to perform maneuver support operations for the echelon it supports. Maneuver support operations integrate the complimentary and reinforcing capabilities of key protection, movement and maneuver, and sustainment functions, tasks, organizations, and systems to enhance freedom of action. The MEB provides protection and mobility to prevent or mitigate effects of hostile action against divisional forces. While the MEB has no direct existing or preceding equivalent units/organizations in today's force structure, it combines functions previously performed by the division rear operations center, division engineer brigade, and other division-level engineer, MP, and chemical assets. MEBs control terrain and potentially facilities and prevent or mitigate hostile actions or weather effects on the protected force. A MEB is a combined arms organization that is task-organized based on mission requirements. It has a combined arms staff and C2 capabilities that suit it for many missions. These brigades typically control combinations of several different types of battalions and separate companies such as MP, CBRN, CA, engineer, EOD, and MI. The MEB may also include AMD units and a tactical combat force (TCF). A TCF is a combat unit that is assigned the mission of defeating level III threats. It is equipped with appropriate tactial enablers and sustainment assets (JP 3-10).

8-101. The mission of the MEB is to conduct maneuver support operations, support area operations, consequence management operations, and stability operations for the supported force. Each of the key tasks that comprise the MEB mission consists of subordinate supporting tasks.

- Conduct maneuver support operations includes these supporting tasks: perform mobility, perform protection, and perform sustainment.
- Conduct support area operations includes: conduct operational area security, conduct response force operations, perform area damage control, conduct terrain management, perform fire support coordination, and conduct airspace management.
- Conduct consequence management includes: respond to CBRN incident, provide support to law enforcement, and conduct post incident response operations.
- Conduct stability operations includes: establish civil security, establish civil control, and restore essential civil services.

8-102. Higher HQ can assign missions for assets assigned or attached to a MEB executed outside its AO, such as CBRN defense and EOD assets. This requires careful coordination between the tasked unit, the MEB HQ, and the BCT for which the mission occurs. The preferred method involves a division HQ cutting a fragmentary order directing the MEB to provide, for example, an EOD capability in direct support of a BCT for a specified period of time. The order authorizes direct liaison between the MEB and the BCT since the MEB will coordinate numerous tactical and sustainment items with the BCT. These items can include movement routes and times, link-up points and times, recognition measures, location of supply points, maintenance collection points, medical facilities, and communications-electronics operating instructions. Alternatively, the division could use the MEB as a force provider. The division could again task-organize the MEB, and the BCT could issue a fragmentary order detaching an EOD team from the MEB and attaching it to the BCT for the duration of the mission. See FM 3-90.31 for more information on the MEB.

## SUSTAINMENT BRIGADE

8-103.   Sustainment brigades are subordinate units of the theater sustainment command. They consolidate functions previously performed by corps and division support commands and area support groups into a single echelon; and they provide C2 of the full range of logistics operations.  Under certain factors of METT-TC, a sustainment brigade could be placed OPCON to a division HQ for a specified operation such as an exploitation or a pursuit operation. However, a division HQ does not routinely have a command relationship with supporting sustainment brigades.

8-104.   Sustainment brigades and their subordinate units are rarely assigned an AO. Their staffs are not configured to perform the standard responsibilities of having an AO. Such responsibilities include:

- Terrain management.
- Movement control.
- Clearance of fires.
- Security operations.
- Stability operations.
- Personnel recovery.
- Reconnaissance and surveillance (to include MI discipline collection).
- Environmental management.

8-105.   However, sustainment brigades have self-protection capabilities, and their commander can be assigned base and base cluster commander responsibilities within an AO assigned to either a BCT or a MEB.

8-106.   The sustainment function consists of related tasks and systems that provide support and services to ensure freedom of action, extend operational reach, and prolong endurance. It includes providing support to BCTs operating in a specified AO. It encompasses the provisioning of personnel services, logistics, health service support, and other support required to sustain combat power. All sustainment brigades have the same general responsibilities: to conduct sustainment operations in an assigned support area. A sustainment brigade supporting a division or joint task force provides sustainment and distribution support to its supported units.

8-107.   During operations, divisions establish a tempo of operations that balances combat and sustainment. This combines mission staging operations and replenishment operations to sustain forces. Two types of rapid replenishment operations complement mission staging operations: sustainment replenishment operations and combat replenishment operations. The BCT S-4 and BSB commander and staff collaborate directly with the supporting sustainment brigade to accomplish replenishment missions. See Chapter 9 for information concerning BCT sustainment operations, and FMI 4-93.2 for more information about the sustainment brigade.

# OTHER BRIGADES AND UNITS

8-108.   A mix of functional brigades and units also share the battlefield with the BCT. These functional brigades usually are assigned or attached to theater-level commands. Examples include MP, engineer, AMD, signal, medical, CBRN defense, and CA. Functional brigades may be attached or OPCON to the Army Force HQ, normally a division. They may also be placed under OPCON of the joint force land component commander. Normally theater-level C2 organizations will augment these functional brigades if they are operating directly under a joint force commander (JFC) and not as part of theater Army.

8-109.   Functional brigades may be allocated to a joint force land component, corps, or division to support the force as a whole or to carry out a particular task. For example, in addition to a MEB, a division might receive a MP brigade HQ and several MP battalions to control dislocated civilians and handle detainees.

## EXPEDITIONARY SIGNAL BATTALION

8-110.   An expeditionary signal battalion plans, engineers, installs, operates, maintains, and defends a minimum of 30 command, control, communications, computers, and information technology nodes. This design supports combatant command, theater Army, JTF, and joint force land component commanders. Additionally, it supports theater-level functional brigades and subordinate battalions regardless of their location in a joint operations area (JOA). See FMI 6-02.45 for more information on expeditionary signal battalions.

## CIVIL AFFAIRS BRIGADE

8-111. The CA brigade functions as the regionally focused, expeditionary, operational-level CA capability that supports the joint task force and its BCTs. It accomplishes its mission by attaching subordinate elements to the BCT. A CA brigade's capabilities include:

- Serve as the BCT commander's senior CA adviser.
- Establish an operational-level CA operations center to support tactical civil-military operations centers in the BCT AO, and facilitate collaboration across multiple brigade AOs.
- Serve as a mechanism for CA coordination to produce focused civil inputs to the COP.
- Plan, coordinate, and enable stability operations with the host nation, international organizations, NGOs, and other government agencies that are focused on the regional to national levels of civil governments. This serves as a CMO staff channel for BCT requirements.
- Provide C2 for subordinate CA battalions and companies that support the BCT.
- Provide cross-cultural communications, limited linguistic capability, and advice to the commander on cultural influences in the area of responsibility.
- Provide the capability to establish the core of a joint CMO task force that can support BCTs.
- Provide the ability to assess, develop, obtain resourcing, and manage operational-level humanitarian assistance, and stability operations spending implementation strategy. (This function requires a dedicated contracting officer and financial management officer.)
- Provide provincial- to national-level civil liaison team capability.
- Provide the civil-information management cell as the focal point for operational-level consolidation and analysis of civil information; developing operational-level civil inputs to the COP with the brigade CA operations center; and linking civil information to the appropriate military and civil systems via geo-spatially referenced data.
- Provide functional specialty cells able to perform an intermediate-level threat assessment to a civil component of the environment at the sub-national level; assess mission planning requirements; and develop, coordinate, and synchronize resources to meet the immediate need in four of the six functional specialties: health and welfare, rule of law, infrastructure, and governance.

8-112. See FM 3-05.40 for more information about CA.

## SECTION X – KEY JOINT ASSETS, FORCES, AND CONSIDERATIONS

8-113. The BCT conducts integrated strike, maneuver, and IO with joint and interagency ground, space, maritime, air, and SOF teams. Such teaming multiplies the combat power of each component enormously, deprives the enemy of the freedom to focus his own efforts, overloads his planning and coordination mechanisms, and compels him to expose his forces to new threats in the effort to evade others. This section provides a discussion of possible military joint teams.

## SPECIAL OPERATIONS FORCES

8-114. Past and ongoing operations have demonstrated that BCTs continue to work in the same tactical space with SOF. In many situations, SOF precede the arrival of the BCTs in operational areas, and constitute a valuable source of intelligence and networking, particularly within the context of irregular warfare.

8-115. The Department of Defense has designated the following as SOF:

- U.S. Army (both active and reserve components). Special forces (SF), Ranger, Army special operations aviation (RSOA), MISO, and CA (FM 3-05).
- U.S. Navy (both active and reserve components). Sea-air-land team (SEAL), SEAL delivery vehicle team, and special boat team units.
- U.S. Air Force (both active and reserve components). Special operations (SO) flying (does not include USAF rescue/combat search and rescue units), special tactics, combat weather, and foreign internal defense (FID) units.
- Conventional units conducting or supporting SO. Designated SOF are the force of choice for the conduct of SO. However, under certain circumstances, conventional forces may be tasked to conduct limited SO on a mission-specific, case-by-case basis (JP 3-05).

8-116. SOF primary tasks include unconventional warfare, foreign internal defense, special reconnaissance, direct action, and counterterrorism. Other duties include combat search and rescue, security assistance, peacekeeping, humanitarian assistance, humanitarian de-mining, counter-proliferation, MISO, and counterdrug operations.

# AIR FORCE SUPPORT

8-117. Air Force tactical air control parties (TACP) are provided to Army maneuver unit HQ from battalion through corps. TACPs advise the commander and staff on the capabilities, limitations, and employment of airpower. Each TACP is a primary point-of-contact to coordinate that echelon's preplanned and immediate air requests and to assist in coordinating air support missions (JP 3-09.3).

8-118. In the TACP, the air mobility liaison officer is the primary adviser on using airlift resources. This officer is trained to control airlift assets in support of ground troops, and operate the airlift advance notification and coordination net.

8-119. The air support operations center co-locates with the main CP when the senior fires cell is in the corps HQ. The senior fires cell directs and monitors fires in the affected AO. The air support operations center commander recommends the location based on the factors of METT-TC.

8-120. The TACP at the main CP is the Air Force element in the division. This TACP is organized as an air execution cell capable of requesting and executing type 2 or 3 control of CAS missions. Manning depends on the situation, but at a minimum, includes an air liaison officer (ALO) and a joint terminal attack controller. Air Force weather and intelligence support may also be incorporated into this TACP.

8-121. The TACP element locates in or adjacent to the fires cell, and provides airpower advice and execution support to the division. Specific air component planning and execution roles include the following:

- Execute airpower according to joint force air component commander's guidance and the division commander's priority, timing, and desired effects within the division AO.
- Inform the commander and staff of the capabilities and limitations of airpower.
- Accomplish training and mission rehearsal under anticipated operational conditions with the USAF and other Service counterparts.
- Exercise OPCON or tactical control of all joint terminal attack controllers operating in the division AO.
- Plan, prepare for, execute, and assess airpower (e.g., CAS, air intelligence, and SEAD) operating in the division AO out to the fire support coordination line.
- Prioritize, coordinate, and deconflict airpower executing missions in the division AO according to the division commander's priorities.
- Provide applicable updates to the COP for air assets tasked to support ground operations.
- Prevent fratricide through situational awareness of the COP and fire support coordinating measures (FSCM).
- Ensure all subordinate TACPs and joint tactical air controllers know and understand joint operations area (JOA) rules of engagement (ROE).

- Deconflict both air and ground assets by monitoring the COP of both friendly and enemy forces reported by intelligence and collaborative tools linked to other C2 units.

- Provide timely and efficient processing of air support requests through collaborative tools and secure communications.

- Provide fast reaction to immediate air support requests, control kill box operations, and integrate and coordinate air support missions such as reconnaissance, electronic warfare, and airlift in the division AO.

- Forward battle damage assessment and aerial weapons effects reports to the air support operations center (ASOC).

## NAVY AND MARINE FORCES

8-122. Navy and Marine Corps forces can project massed offensive and defensive combat power from the sea, at the time and place of their choosing. They can deter potential adversaries through the ability to preempt or interdict aggressive action (Naval Doctrine Publication [NDP] 1).

8-123. The Marine air-ground task force (MAGTF) is a term the U.S. Marine Corps uses to describe the principal organization for all missions across the spectrum of conflict. MAGTF is a balanced air-ground, combined arms task organization of Marine Corps forces, structured to accomplish a specific mission, under a single commander. A Marine expeditionary force (MEF) is the largest type of a MAGTF. Each MEF consists of a division as the ground combat element, an aircraft wing as the aviation combat element, and a logistics group as the logistics combat element (Marine Corps Doctrine Publication [MCDP] 3).

8-124. A Marine expeditionary brigade is capable of conducting missions across the spectrum of conflict and varies in size. It is built around a reinforced Infantry regiment, a composite Marine aircraft group, and a brigade service support group. The Marine expeditionary brigade is task-organized to meet the requirements of a specific situation. It can function alone, as part of a JTF, or as the lead echelon of the MEF.

8-125. A Marine expeditionary unit (MEU) is the smallest MAGTF. Each MEU is an expeditionary quick reaction force, deployed and ready for immediate response to any crisis. Usually, the MEU is built around a reinforced Marine Infantry battalion with a composite helicopter squadron as the aviation combat element, a battalion-sized logistics combat element, and a command element.

8-126. Troop strength is about 2,200, commanded by a colonel, and is deployed from an amphibious assault ship. It task organizes as:

- The MEU's command element , which includes the MEU commander and his supporting staff, and provides C2. It includes specialized detachments for naval gunfire, reconnaissance, surveillance, specialized communications, radio reconnaissance (signals intelligence), electronic warfare, intelligence and counterintelligence, and PA missions. The overall strength is about 200 Marines and sailors.

- The MEU's ground combat element consists of the battalion landing team; an Infantry battalion reinforced with an artillery battery, amphibious assault vehicle platoon, combat engineer platoon, light armored reconnaissance company, tank platoon, reconnaissance platoon; and other units as the mission and circumstances require. The total strength is approximately 1,200 members.

- The MEU's aviation combat element (ACE) is a Marine composite squadron (reinforced). It consists of a medium/heavy helicopter squadron augmented with three other types of helicopters; one detachment of amphibious flight-deck-capable jets; and a Marine air control group detachment with air traffic control, direct air support, and antiaircraft assets.

- The MEU's logistics combat element (LCE) contains all the logistics specialists and equipment necessary for the unit to support itself for 15 days in an austere expeditionary environment. It includes service support, medical, dental, maintenance, transportation, explosive ordinance disposal, MP, utilities production and distribution, bulk fuels, and other technical experts. It consists of approximately 300 Marines and Sailors. This element may be referred to as the MEU service support group.

## SECTION XI – MULTINATIONAL CONSIDERATIONS

8-127.  In any large-scale operation, the BCT usually operates with multinational partners, either allied or host nation. Many considerations associated with such coalitions exist at the operational level. Here, U.S. commanders must overcome significant challenges in interoperability to integrate multinational forces effectively within the joint campaign force structure and design. To the extent that multinational forces are integrated into tactical operations at the division level and below, tactical commanders must also actively work to ensure effective C2, communications, and information sharing. The multinational force and BCT HQ may be required to exchange liaison elements, including information systems.

This page intentionally left blank.

# Chapter 9

# Sustainment Operations

The brigade support battalion (BSB) is the core of sustainment for the brigade combat team (BCT). The BSB is organic to the BCT and consists of functional and multifunctional companies assigned to provide support to the BCT. The BSB of the Stryker Brigade Combat Team (SBCT) task-organizes to provide support to individual Infantry battalions. This chapter provides an overview of sustainment operations within the BCT and sustainment planning considerations for BCT full spectrum operations.

## SECTION I – BRIGADE COMBAT TEAM SUSTAINMENT

9-1.   BCTs are organized with the self-sustainment capability for up to 72 hours of combat. Beyond 72 hours, sustainment organizations at the division and corps levels are required to conduct replenishment of the BCT's combat loads. That replenishment is a function of the higher headquarters sustainment brigade(s).

## SUSTAINMENT BRIGADES

9-2.   The sustainment brigade is a scalable, adjustable, networked brigade comprised of a headquarters, and both functional and multifunctional subordinate sustainment units. The theater sustainment command uses sustainment brigades to provide operational-level support to theater armies. All sustainment brigades provide area support, although the specific tasks they are assigned may differ (Figure 9-1).

9-3.   The combat sustainment support battalions of the sustainment brigade are the base organization from which sustainment units are task-organized for various operations. The combat sustainment support battalion subordinate elements consist of functional companies that provide supplies and services, ammunition, fuel, transportation, and maintenance. Additionally, personnel and financial management units can be assigned to, or administratively controlled by, the combat sustainment support battalions to perform essential human resources and finance operations. The combat sustainment support battalions provide the distribution link between theater aerial/sea ports of debarkation and the BCT's BSB. The structure includes cargo transfer and movement control assets, performing the function of transporting commodities to and from the BCT BSB, and to/from repairing or storage facilities at the theater base. Its function is to ensure and maintain the flow of replenishment using expeditionary support packages, including retrograde of unserviceable components, end items and supplies.

9-4.   Battalion medical platoons and the brigade support medical company provide health service support and force health protection to BCTs. The theater Army has a medical command (deployment support) (MEDCOM [DS]) for command and control of all medical units in a theater of operations at echelons above brigade. The MEDCOM (DS) provides subordinate medical organizations that operate under the medical brigade and/or multifunctional medical battalion. The medical brigade provides a scalable expeditionary medical capability for assigned and attached medical organizations that are task-organized to support BCTs and echelons above brigade. The multifunctional medical battalion also provides medical C2, administrative assistance, logistical support, and technical supervision for assigned and attached companies and detachments. The multifunctional medical battalion is assigned to the MEDCOM (DS) or medical brigade. The combat support hospital is also a battalion-size element assigned to the MEDCOM (DS) or medical brigade.

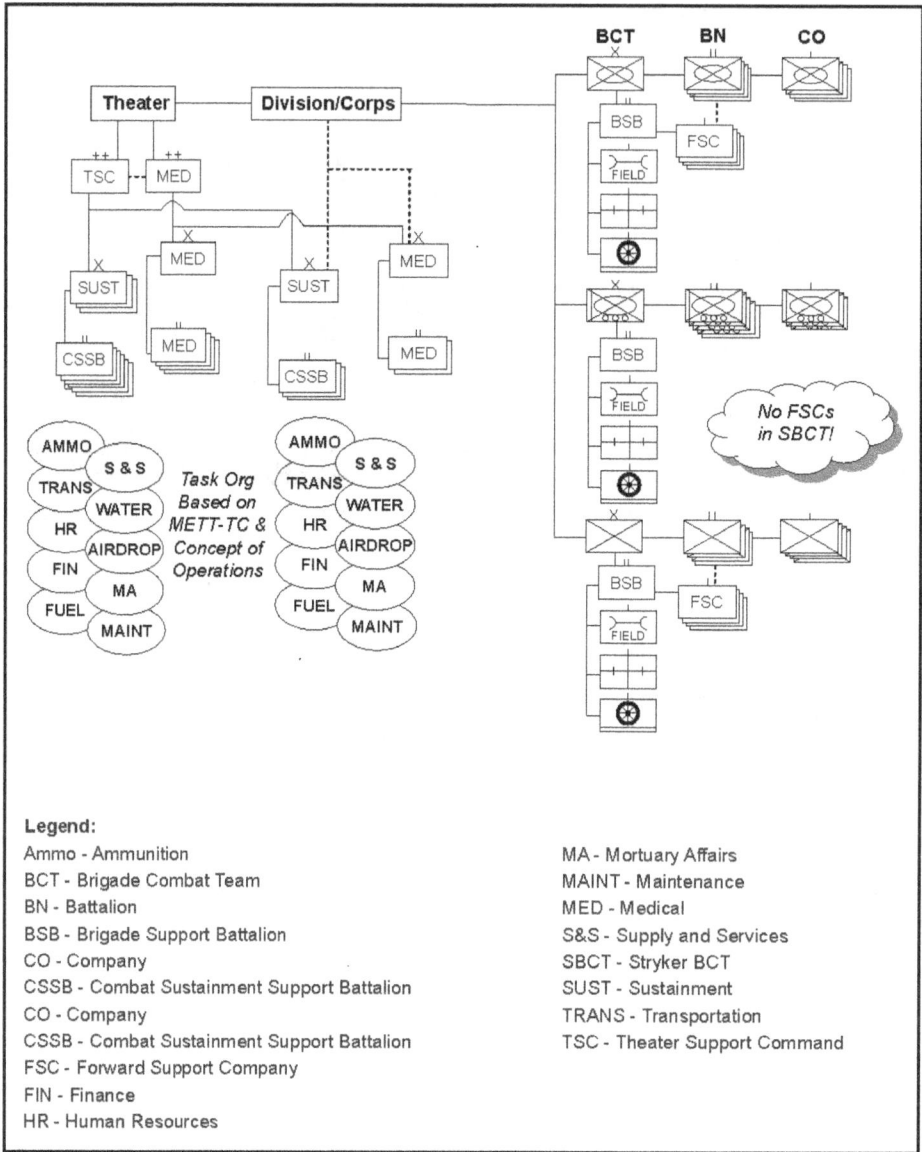

**Figure 9-1. Sustainment organizations supporting the BCT**

## SUSTAINMENT FUNCTIONS

9-5.   The sustainment warfighting function provides support under three categories: logistics, personnel services, and health service support. Table 9-1 provides the functional elements that are found in each of the sustainment categories applicable to the BCT. The BCT's organic sustainment capabilities require augmentation to provide some of these functions and services.

Table 9-1. BCT sustainment categories and elements

| Logistics | Personnel Services | Health Service Support |
|---|---|---|
| Maintenance | Human resources support | Organic and area medical support |
| Transportation | Financial management | Behavioral health/neuropsychiatric treatment (treatment aspects) |
| Supply | Legal support | Hospitalization |
| Field services | Religious support | Dental care (treatment aspects) |
| Distribution | | Clinical laboratory services |
| Contracting | | Treatment of CBRN patients |
| General engineering support | | Medical evacuation |
| | | Medical logistics |

Source: FM 3-0

# SUSTAINMENT STAFF AND ORGANIZATIONS

9-6. The following paragraphs provide additional information on the BCT sustainment staff and organizations described in Chapter 1. It also describes the support that elements other than the BSB provide.

## BRIGADE COMBAT TEAM PERSONNEL AND ADMINISTRATION SECTION

9-7. The BCT personnel and administration section personnel staff officer (S-1) is responsible for planning, coordinating, and executing human resources support. He is responsible for manning the BCT and subordinate units while maintaining personnel accountability of all personnel assigned and attached to the BCT. The S-1 provides essential personnel services (awards, evaluations, ID cards, etc.), conducts postal support, manages the casualty reporting system, and provides administrative support for BCT units. The S-1 coordinates external human resources support through the higher headquarters (G-1) and the HR operations branch within the supporting sustainment brigade. The S-1 section relies on automated personnel systems for updating personnel management information. The S-1 coordinates Army command interest programs along with coordinating financial management and legal support for the BCT. Although the S-1 coordinates the staff efforts of the BCT chaplain and BCT surgeon, they generally receive their guidance from the BCT executive officer (XO). The S-1 is also the staff point of contact for inspector general and morale support activities. See FM 1-0 for additional information about S-1 roles and responsibilities.

### Brigade Combat Team Unit Ministry Team

9-8. The BCT unit ministry team (UMT) is responsible for organizing the efforts of UMTs that work for subordinate commanders. The BCT UMT must ensure there is religious support to all Soldiers in the BCT AO. Often, companies or detachments are attached to the BCT without UMT support. Members of other services and authorized civilians may require area support. The BCT UMT is responsible for the professional oversight of the battalion UMTs.

### Brigade Combat Team Surgeon Section

9-9. The BCT surgeon is a special staff officer responsible for Army Health System (AHS) support in the BCT. The BCT surgeon exercises technical control over medical activities in the BCT. He provides staff oversight and supervision for AHS support and keeps the BCT commander informed of the health of the command. The BCT surgeon's section ensures timely planning, integration, and synchronization of AHS support with the BCT maneuver plan. The BCT surgeon's section coordinates with the brigade support medical company; the battalion medical platoons and sections; the BSB medical operations section; and other staff elements to ensure that Soldiers receive complete and comprehensive medical support.

## BRIGADE OPERATIONAL LAW TEAM

9-10. The BCT judge advocate is the special staff officer responsible for operational and administrative law support to the BCT. He is the BCT commander's personal legal advisor. The brigade operational law team provides administrative legal services to BCT and battalion S-1s, and may provide legal advice on host nation support (HNS) issues to the BCT logistics staff officer (S-4). The operational law team members provide the BCT staff with immediate access to the legal expertise they need to prevail in an increasingly complex and legally intensive operational environment.

## BRIGADE COMBAT TEAM COMMUNICATIONS SECTION

9-11. The BCT communications section signal staff officer (S-6) is responsible for maintaining selected components of the BCT's command and control (C2) information system and network operations. The S-4 and S-6 must coordinate to ensure there are no gaps in the maintenance system for communications security (COMSEC) equipment, computers, and other specialized equipment.

# SUSTAINMENT SUPPORT AREAS

9-12. A support area is a designated area in which sustainment elements, some staff elements, and other elements locate to support a unit. The BCT S-4, BCT S-3, and BSB S-3 coordinate the location of the BCT sustainment support areas. The BCT S-3 coordinates with the BCT subordinate battalions to ensure that they have adequate space to position their sustainment units. Types of support areas include:

* Battalion and company trains.
* Brigade support area (BSA).

## TRAINS

9-13. Trains are a grouping of unit personnel, vehicles, and equipment to provide sustainment. It is the basic sustainment tactical organization. Maneuver battalions use trains to array their subordinate sustainment elements including their designated forward support company (FSC). Battalion trains usually are under the control of the battalion S-4, assisted by the battalion S-1. The composition and location of battalion trains varies depending on the number of units attached to, or augmenting, the battalion.

9-14. Battalion trains can be employed in two basic configurations: as unit trains in one location, or as echeloned trains:

* Unit trains at the battalion level are appropriate when the battalion is consolidated, during reconstitution, and during major movements.
* Echeloned trains can be organized into company trains, battalion combat trains, unit maintenance collection point (UMCP), battalion aid station (BAS), or battalion field trains.

### Company Trains

9-15. Company trains provide sustainment for a company during combat operations. Company trains usually include the first sergeant, medical aid/evacuation teams, supply sergeant, and the armorer. Usually, the forward support company provides a field maintenance team with capabilities for maintenance, recovery, and limited combat spares. The supply sergeant can collocate in the combat trains if it facilitates logistics package (LOGPAC) operations. The first sergeant usually directs movement and employment of the company trains; although the company commander may assign the responsibility to the company XO.

### Battalion Trains

9-16. Battalion trains consist of two types: combat trains and field trains.

#### Combat Trains

9-17. The combat trains usually consist of the forward support company and the battalion medical unit. The UMCP should be positioned where recovery vehicles have access, or where major or difficult maintenance is performed. Units must consider the mission enemy terrain and weather, troops and support available, time available, and civil considerations (METT-TC) when locating combat trains in a battalion support area.

*Field Trains*

9-18. Field trains can be located in the BSA and include those assets not located with the combat trains. The field trains can provide direct coordination between the battalion and the BSB. When organized, the field trains usually consist of the elements of the forward support company, battalion headquarters and headquarters company (HHC), battalion S-1, and battalion S-4. The field trains personnel facilitate the coordination and movement of support from the BSB to the battalion.

## Sustainment-Related Command Posts

9-19. The battalion commander may choose to create a combat trains command post (CTCP) or a field trains command post (FTCP) to control administrative and sustainment support. Most of the time, the S-4 is the officer in charge of the CTCP. If constituted, the FTCP could be led by the HHC commander. These command posts (CP) can be organized to accomplish specific logistical tasks:

- Situations that might dictate the need for a CTCP:
  - As part of a BSB forward logistics element operation.
  - During reception, staging, onward movement, and integration operations.
- Situations that might dictate the need for a FTCP:
  - During periods of supply or resupply of major end items.
  - The sustainment elements of the battalion are no longer 100% mobile.

## BRIGADE SUPPORT AREA

9-20. The BSA is the logistical, personnel, and administrative hub of the BCT. It consists of BSB, but could also include a BCT alternate command post (if formed), battalion field trains, brigade special troops battalion units, signal assets, and other sustainment units from higher headquarters (HQ). The BCT operations staff officer (S-3), with the BCT S-4 and the BSB, determines the location of the BSA. The BSA should be located so that support to the BCT can be maintained, but does not interfere with the tactical movement of BCT units, or with units that must pass through the BCT area. The BSA's size varies with terrain; however, an area 4 km to 7 km in diameter is a planning guide. Usually the BSA is on a main supply route (MSR) and ideally is out of the range of the enemy's medium artillery. The BSA should be positioned away from the enemy's likely avenues of approach and entry points into the BCT's main battle area (MBA). The BSB commander is responsible for the command and control of all units in the BSA for security and terrain management

9-21. Usually the BCT S-4 coordinates the BCT main CP's sustainment cell, which contains the BCT S-4, BCT S-1, BCT surgeon section, and the BCT UMT. The BCT commander can create an alternate CPs for sustainment, should the administrative and logistics presence in the main CP become too large. The brigade special troops battalion CPs may be able to host the sustainment cell in the BSA if communications links are adequate.

## Locations for Support Areas

9-22. The trains should not be considered a permanent or stationary support area (Figure 9-2). The trains must be mobile to support the battalion when it is moving, and should change locations frequently, depending on available time and terrain. The trains changes locations for the following reasons:

- Change of mission.
- Change of unit AOs.
- To avoid detection caused by heavy use or traffic in the area.
- When area becomes worn by heavy use (e.g., wet and muddy conditions).
- Security becomes lax or complacent due to familiarity.

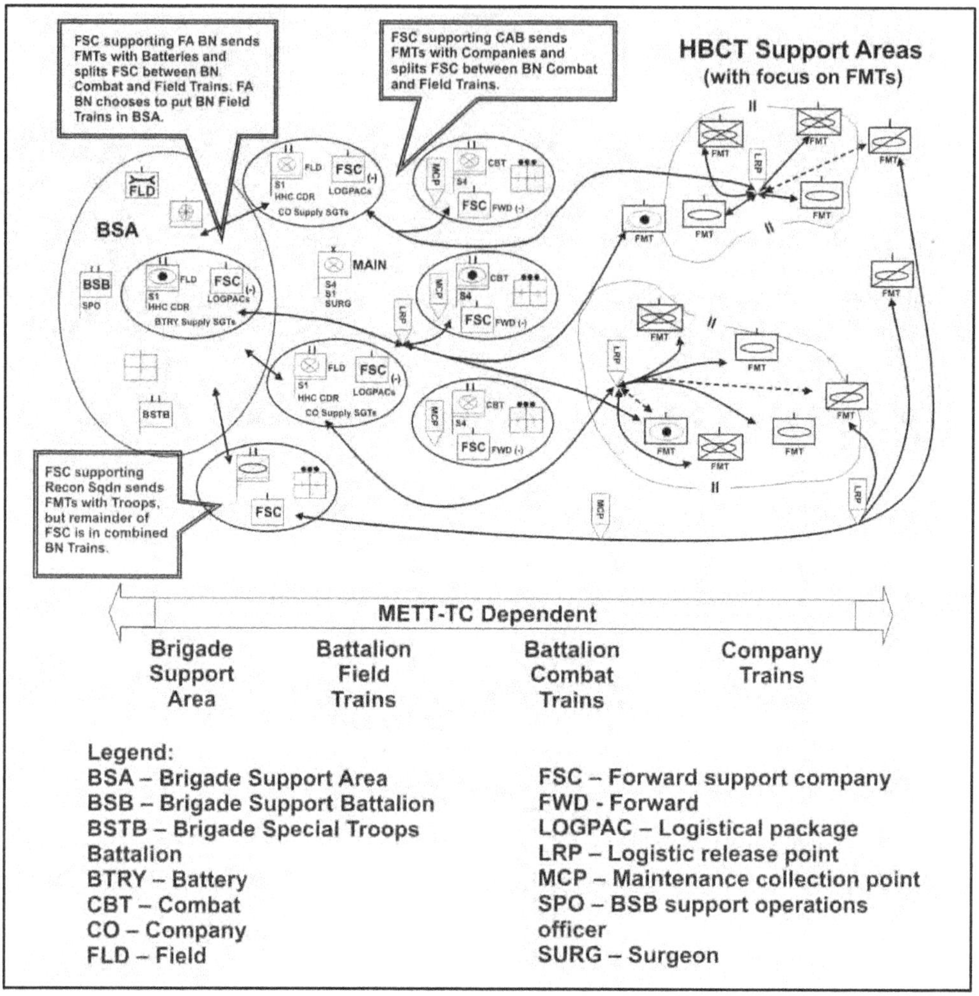

**Figure 9-2. Support unit locations**

9-23. All support areas have many similarities, including:

- Cover and concealment (natural terrain or manmade structures).
- Room for dispersion.
- Level, firm ground to support vehicle traffic and sustainment operations.
- Suitable helicopter landing site (remember to mark the landing site).
- Good road or trail networks:
  - Good routes in and out of the area (preferably separate routes going in and going out).
  - Access to lateral routes.
  - Positioned along or good access to the MSR.
  - Positioned away from likely enemy avenues of approach.

## Security of Support Areas

9-24. Sustainment elements must organize and prepare to defend themselves against ground or air attacks. They often occupy areas that maneuver elements of the BCT have secured. The security of the trains at each echelon is the responsibility of the individual in charge of the trains. The best defense is to avoid detection. The following activities help to ensure trains security:

- Select good trains sites that use available cover, concealment, and camouflage.
- Use movement and positioning discipline as well as noise and light discipline to prevent detection.
- Establish a perimeter defense.
- Establish observation posts and patrols.
- Position weapons (small arms, machine guns, and antitank weapons) for self-defense.
- Plan mutually supporting positions to dominate likely avenues of approach.
- Prepare a fire plan and make sector sketches.
- Identify sectors of fires.
- Emplace target reference points (TRP) to control fires and for use of indirect fires.
- Integrate available combat vehicles within the trains (i.e., vehicles awaiting maintenance or personnel) into the plan, and adjust the plan when vehicles depart.
- Conduct rehearsals.
- Establish rest plans.
- Identify an alarm or warning system that would enable rapid execution of the defense plan without further guidance; the alarm, warning system, and defense plan are usually included in the standard operating procedure (SOP).
- Designate a reaction force. Ensure the force is equipped to perform its mission. The ready reaction force must be well rehearsed or briefed on—
  - Unit assembly.
  - Friendly and threat force recognition.
  - Actions on contact.

## Supply Routes

9-25. The BCT S-4, in coordination with the BSB support operations officer and BCT S-3, select supply routes between support areas. MSRs are routes designated within the BCT's AO upon which the bulk of sustainment traffic flows in support of operations. An MSR is selected based on the terrain, friendly disposition, enemy situation, and scheme of maneuver. Alternate supply routes are planned in the event that an MSR is interdicted by the enemy or becomes too congested. In the event of chemical, biological, radiological, and nuclear (CBRN) contamination, either the primary or the alternate MSR can be designated as the "dirty MSR" to handle contaminated traffic. Alternate supply routes should meet the same criteria as the MSR. Military police (MP) may assist with regulating traffic, and engineer units, if available, could maintain routes. Security of supply routes in a noncontiguous environment might require the BCT commander to commit non-sustainment resources.

9-26. Route considerations include:

- Location and planned scheme of maneuver for subordinate units.
- Location and planned movements of other units moving through the BCT's AO.
- Route characteristics such as route classification, width, obstructions, steep slopes, sharp curves, and type of roadway surface.
- Two-way, all-weather trafficability.
- Classification of bridges and culverts.
- Requirements for traffic control such as at choke points, congested areas, confusing intersections, or along built-up areas.
- Number and locations of crossover routes from the MSR to alternate supply routes.
- Requirements for repair, upgrade, or maintenance of the route, fording sites, and bridges.
- Route vulnerabilities that must be protected. This can include bridges, fords, built-up areas, and choke points.

- Enemy threats such as air attack, conventional and unconventional tactics, mines, ambushes, and chemical strikes.
- Known or likely locations of enemy penetrations, attacks, chemical strikes, or obstacles.
- Known or potential civilian/refugee movements that must be controlled or monitored.

# MISSION TAILORING OF SUSTAINMENT ASSETS

9-27. Sustainment operations can be tailored in response to changes in tactical requirements. In most cases, the BSB will provide the supplies and services required by the supported unit at a specific point in time (scheduled delivery). For example, a typical day may include distribution to a battalion level distribution point for one customer cluster, to company/battery level for another customer cluster, and all the way to platoon/team level for a third cluster, while the fourth cluster receives no delivery (due to low/no requirements) that particular day.

9-28. Supported unit commanders coordinate through their S-3 and S-4 staffs, according to current unit battle rhythm, to fix the time and place for replenishment operations at a temporarily established point. Assets can be retasked if the situation demands. This approach, executed according to centralized management, optimizes the employment of personnel.

# SYNCHRONIZATION OF BATTLE RHYTHM AND SUSTAINMENT OPERATIONS

9-29. Sustainment operations are fully integrated with the BCT battle rhythm through integrated planning and oversight of ongoing operations. Sustainment and operational planning occurs simultaneously rather than sequentially. Incremental adjustments to either the maneuver or sustainment plan during its execution must be visible to all BCT elements. The sustainment synchronization matrix and sustainment report are both used to initiate and maintain synchronization between operations and sustainment functions.

## SITUATIONAL UNDERSTANDING

9-30. Situational understanding (SU) is the complete understanding by the BCT commander of the friendly situation, the enemy situation (as described by current intelligence), and the sustainment situation using advanced, seamless information technology. Key elements of SU are:

- A common operational picture (COP) that enables the BCT maneuver and sustainment commanders, along with the S-4s and S-1s to view the same data in near real time fostering unity of command and unity of effort.
- An integrated, seamless information network bringing together in-transit visibility, unit requirements, and COP in near real time and sharing the information across sustainment functions and infrastructure while allowing the exchange of large volumes of information across platforms.
- Timely and accurate asset visibility information that enables the distribution of assets on time, maintaining the critical confidence in the distribution system. Visibility begins at the point where resources start their movement to the AO. In-transit visibility uses advanced automation, information, and communications capabilities to track cargo and personnel while en route.
- In addition to COP, liaison officers (LNO) often are embedded at the maneuver brigade staff to pass current commander intent and mission changes to sustainment elements.
- The BCT S-4 and S-1 contribute to the BCT commander's SU by continually providing assessments of BCT sustainment operations. These running estimates present recommendations to mitigate threats and capitalize on opportunities.

## FUSION OF SUSTAINMENT AND MANEUVER SITUATIONAL UNDERSTANDING

9-31. Effective sustainment operations by the BSB depend on a high level of SU and COP. SU enables the BSB commander and staff to maintain visibility of current and projected requirements; to synchronize movement and materiel management; and to maintain integrated visibility of transportation and supplies.

Battle Command Sustainment Support System (BCS3), Movement Tracking System , and Force XXI Battle Command Brigade and Below (FBCB2) are some of the fielded systems that the BSB uses to ensure effective SU and logistics support. These systems enable sustainment commanders and staffs to exercise centralized C2, anticipate support requirements, and maximize battlefield distribution.

## REPORTS

9-32. Accurately reporting the sustainment status is essential to keeping units combat ready. SOPs should establish report formats, reporting times, and FM voice brevity codes to keep logistics nets manageable. The FBCB2 system helps lower level commanders automate the sustainment data-gathering process. It does this through logistics situation reports (LOGSITREP), personnel SITREPs, logistical call for support, logistics task order messaging, situational awareness (SA), and task management. This functionality affects the synchronization of all logistics support on the battlefield between the supported and the supporter.

9-33. At the BSB and BCT levels, BCS3 collects sustainment data from various logistics-related standard Army management information systems throughout the BCT. These systems provide actionable logistics information to support sustainment-related decisions. To assist planners, BCS3 has a simulation tool that enables the user to project supply consumption for a given course of action by event or across time. To assist in execution, BCS3 gives the commander the latest available status of critical weapons systems, fuel, ammunition, and personnel. BCS3 also provides a map-centric view of inbound vehicles and/or cargo that are equipped with movement tracking devices.

9-34. Although sustainment planners may have data available from BCS3 and FBCB2 logistics and personnel status messages, they may need to use nonstandard text messages to identify equipment and personnel issues. The sustainment staff must proactively identify and solve sustainment issues by:

- Using FBCB2, BCS3 and other Army Battle Command Systems (ABCS) to maintain sustainment SA.
- Working closely with higher HQ staff to resolve sustainment problems.
- Recommending sustainment priorities that conform to mission requirements.
- Recommending sustainment-related commander's critical information requirement (CCIR).
- Ensuring the commander is apprised of critical sustainment issues.
- Coordinating as required with key automated system operators and managers to assure mission focus and continuity of support.

9-35. The S-6 and the information systems technician must work together to ensure that FBCB2, BCS3 and sustainment standard Army management information systems have interconnectivity. The BCT S-4, BCT S-1, and BSB support operations officer must monitor the health of this system and implement alternate means of reporting as necessary (Figure 9-3).

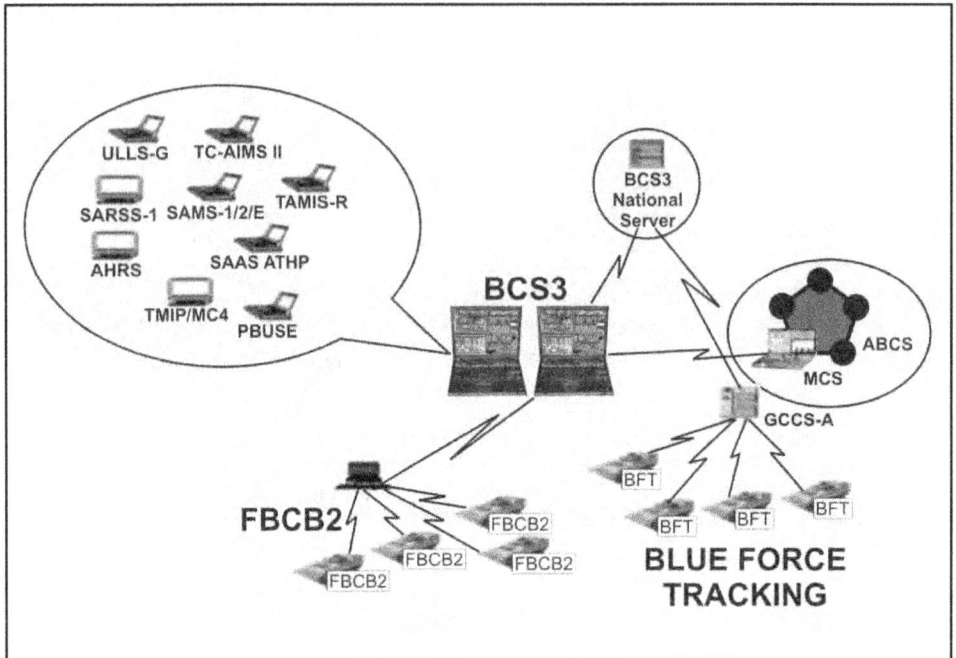

**Figure 9-3. Digital sustainment reporting**

**Medical Reporting**

9-36. The Defense Health Information Management System (DHIMS) and Medical Communications for Combat Casualty Care (MC4) support information management requirements for the BCT surgeon's section and the BCT medical units. The brigade surgeon's section uses BCS3, FBCB2, and DHIMS/MC4 to support mission planning, coordination of orders and subordinate tasks, and to monitor/ensure execution throughout the mission.

9-37. The DHIMS/MC4 is an automated system that links health care providers and medical support providers, in all roles of care, with integrated medical information. The DHIMS/MC4 receives, stores, processes, transmits, and reports medical C2, medical surveillance, casualty movement/tracking, medical treatment, medical situational awareness, and medical logistics data across all roles of care.

# SUSTAINMENT REACH OPERATIONS

9-38. Echelons above brigade (EAB) sustainment reach operations reinforce the BSB. Sustainment reach operations mean using and positioning all-available sustainment assets and capabilities— from the national sustainment base through the Soldier in the field—to support full spectrum operations. The goal of sustainment reach is to reduce the amount of supplies and equipment in the AO in order to sustain combat power more quickly. Reach operations include, but are not limited to, external sources of information and intelligence, sustainment planning and analysis conducted outside the AO, telemedicine, and other temporarily required capabilities. The BSB exploits regionally available resources through joint, multinational, host nation, or contract sources for certain bulk supplies and services.

9-39. Operational contract support is potentially a critical part of sustainment reach operations. Because the BCT is a requiring activity, the BCT commander and sustainment staff need to be familiar with operational contract support requirements, roles, and responsibilities. Two key organizations responsible for coordinating contracted support and for providing other national strategic operational contracting support capabilities, are the Army field support brigade (AFSB) and the Army contracting support brigade (CSB).

The AFSB is responsible for coordinating system support contract actions, and contract efforts related to sustainment maintenance. The CSB is responsible for theater support contracting actions. Both the AFSB and CSBs are subordinate commands of the United States Army Materiel Command. See FM 4-92, FMI 4-93.41, and JP 4-10 for more information.

9-40. The support operations officer is the principal staff officer for coordinating sustainment reach operations for supported forces. The support operations section is the key interface between the supported units and the source of support. The support operations officer advises the commander on support requirements versus available support assets. Sustainment reach operations involve risk analysis. The commander ultimately decides which support capabilities must be located within the AO and which must be provided by a reach operation. METT-TC factors and command judgment determine this ratio. The support operations officer also determines which reach resources can be directly coordinated and which must be passed to the next higher support level for coordination. If resources must be contracted, the support operations staff (external support) or the BCT S-4 (internal support) prepares an acquisition package and submits it to the approving official(s). Once approved and funded, the support operations staff passes the package to the supporting contracting officer who works within the context of the Army/joint contracting framework for the operation.

9-41. The support operations officer continually updates sustainment reach requirements based on the sustainment plan. Planning is the process of gathering data regarding pertinent battlefield components, analyzing their impact on the sustainment estimate, and integrating them into tactical planning so that support actions are synchronized with maneuver. It is a conscious effort to identify and assess those factors that facilitate, inhibit, or deny support to combat forces. Using sustainment planning, the BSB commander chooses from among a number of alternatives and recommends those that best support the maneuver commander's priorities and missions.

9-42. HNS is one of the more commonly used sources of sustainment reach support. Host nations provide support to Army forces and organizations located in or transiting through host nation territory. This support can include assistance in almost every aspect required to support military operations within the AO, including both civil and military assistance. Planners must consider that HNS meets local, and not necessarily U.S., standards. Commanders must consider additional support requirements generated by using HNS. For example, HNS provision of potable water may mean bulk water from a desalinization plant, not bottled water, which increases requirements for tankers and a distribution system. Using HNS should not degrade required U.S. unilateral capability. A theater support command includes an HNS directorate. This directorate exercises staff supervision over Army forces' HNS functions, and recommends allocation of resources to EAB support requirements.

9-43. When planning sustainment reach operations, commanders must conduct a thorough risk analysis of the mission. Reach operations are vulnerable and highly susceptible to many factors, including U.S. Army commanders' distrust of non-U.S. support, changes to the political situation, direct or indirect terrorist activity, local labor union activities, language differences, quality assurance/quality control challenges, compatibility issues, and legal issues. Since reach operations involve DOD civilians, contractors, and joint and multinational forces, protection operations become paramount to mission accomplishment. Single individuals within an agency can carry out terrorist operations. For example, an employee working for a contractor could contaminate food and water sources with small amounts of a biological agent. This could be even more deadly than a truck packed with explosives. Commanders must implement countermeasures to prevent reach operations from becoming an opportunity for terrorist action against U.S. forces.

9-44. Commanders must carefully weigh the benefits of reach operations against protection requirements, especially when using reach assets that are not U.S. military forces. The ability and tendency of our enemies to use asymmetrical force against U.S. forces increases the inherent risks of some reach operations. If local hostilities escalate, support provided by civilians or contractors may be disrupted also. Commanders must consider potential risks and develop detailed plans for compensating for sudden variances in reach support. The limited size and capability of the BSB is further stressed by protection requirements of the forward operating base (when positioned as a stand-alone base versus as part of a base cluster) and during convoy operations.

9-45. Protecting contractors and government civilians on the battlefield is the commander's responsibility. When contractors perform in potentially hostile areas, the supported military forces must protect their operations and personnel. Contractors are subject to the same threat as Soldiers, but they cannot be required to perform protection functions. Contracted personnel do retain the inherent right to self-defense. Commanders should assess whether or not contractor support is vital enough to warrant a diversion of forces to contractor security duties.

9-46. Units or activities requiring support in high-risk contingencies must carefully list the required services and specify the working conditions so the contractors know what they are expected to deliver. The cost of the contract may increase substantially if the contractor is asked to perform under dangerous conditions. Additionally, contractors may be willing to accept more risk if the Army meets specified security requirements such as escorting, training, or providing site security to ensure their safety. Commanders must accept responsibility for the security of contractor personnel when they use contracted support.

## SECTION II – SUSTAINMENT PLANNING

9-47. The lead planner for sustainment in the BCT is the BCT S-4, assisted by the BCT S-1, the BCT surgeon, and the BSB support operations officer. Representatives from these and other sections form a sustainment planning cell at the BCT main CP to ensure sustainment plans are fully integrated into all operations planning. The unit SOP should be the basis for sustainment operations, with planning conducted to determine specific requirements and to prepare for contingencies. BCT and subordinate unit orders should address only specific support requirements for the operation and any deviations from SOP.

9-48. Although the sustainment planners at the BCT main CP control and coordinate sustainment for specific BCT operations, routine operations usually are planned in the BSA. The BCT S-1 may have representatives at or near the BSB to handle various personnel functions (e.g., replacement or monitor casualty operations). The BCT S-4 might choose to locate the property book or movement sections with the BSB support operations section. The BSB commander and staff, along with their planning responsibilities are also responsible for executing the BCT sustainment operations.

9-49. To provide effective support, sustainment planners and operators must understand the mission statement, intent, and concept of the operation. These will lead to developing a concept of support that the BCT operation order (OPORD) provides. The BCT S-4 is responsible for producing the sustainment paragraph and annexes of the OPORD, which should include the following:

- Commander's priorities.
- Class (CL) III/ V resupply during the mission, if necessary.
- Movement criteria.
- Type and quantities of support required.
- Priority of support, by type and unit.
- Sustainment overlay.
- Supply routes.
- Logistic release points.
- Casualty evacuation points.
- Maintenance collection points.
- Operational contract support.

## OPERATIONAL CONTRACT SUPPORT

9-50. Routine BCT S-4/BSB support operations contract support-related tasks include:

- **Planning.** The BCT staff, in coordination with other BCT staff members, the support sustainment command and supporting CSB, integrates contract support into tactical planning actions based on their higher headquarters contract support integration plan and other policy guidance.

- **Developing in-theater requirements.** Usually the BCT is a requiring activity for operational contract support requests, such as locally purchased commodities and simple services. As such, its logistics staff develops acquisition-ready requirement packets, detailing the required support, for submission to the supporting contracting activity. The packets include a detailed performance work statement for service requirements or detailed item descriptions for a commodity requirement. In addition to the performance work statement, these packets include an independent cost estimate and DA Form 3953, *Purchase Request and Commitment*, and letter of justification per local command guidance. Finally, the BCT must be prepared to support an acquisition review board, which approves and sets priorities on high demand and special command interest contract support actions, again as required by local policy.
- **Assisting the contract management and contract quality control process.** BCTs are responsible for providing contracting officer representatives for all service contracts, and a receiving official for commodity contracts. These unit-nominated individuals are critically important to ensure that contractors provide the contracted service or item in accordance with the stipulations of the contract.
- **Contract management.** In coordination with the other BCT staff, supporting sustainment command, CSB and AFSB, ensure that contractor personnel who perform services within the BCT's AO are properly supported as per the stipulation of the contract, and are integrated into the local force protection/security plans in accordance with local command guidance.

9-51. BCT commanders requiring contracted supplies, services and other commercial support must identify and provide personnel to serve as contracting office representative (COR) as required by the supporting contracting officer(s). COR, as subject matter experts in the support provided by the contract, monitor contractor performance within the scope of their delegation to ensure the BCT receives the support they required in their statement of work. Contracting officers appoint, issue delegations to and manage the contract administration work of COR. Only a contracting officer may appoint or relieve a COR of their duties as a COR.

9-52. The remainder of this section provides a description of the concepts that apply to planning sustainment support within the BCT. It also describes how the BCT organizes the sustainment staff and organizations during full spectrum operations.

# CONCEPT OF SUPPORT

9-53. The concept of support (paragraph 4 of an OPORD) establishes priorities of support (by phase or before, during, and after) for the operation. The BCT commander sets these priorities for each level in his intent statement and the concept of the operation. Priorities include such things as personnel replacements; maintenance and evacuation, by unit and by system (aviation and surface systems would be given separate priorities); fuel and/or ammunition; road network use by unit and/or commodity; and any resource subject to competing demands or constraints.

9-54. To establish the concept of support, the BCT sustainment planners must know:

- Subordinate units' missions.
- Times missions are to occur.
- End states.
- BCT scheme of movement and maneuver.
- Timing of critical events.

# SUPPORTING OFFENSIVE OPERATIONS

9-55. Sustainment in the offense is characterized by high-intensity operations that require anticipatory support as far forward as possible. Commanders and staffs ensure adequate support for continuing the momentum of the operation as they plan and synchronize offensive operations. Plans should include agile and flexible sustainment capabilities to follow exploiting forces and continue support (FM 4-0).

9-56. The following sustainment techniques and considerations apply to offensive planning:

- Plan for dealing with threats to sustainment units from bypassed enemy forces in a fluid, non-contiguous AO.
- Recover damaged vehicles only to the main supply route for further recovery or evacuation.
- Pre-stock essential supplies forward to minimize interruption to lines of communications.
- Plan for increased consumption of petroleum, oils, and lubricants (POL).
- Anticipate increasingly long lines of communications as the offensive moves forward.
- Anticipate poor trafficability for sustainment vehicles across fought-over terrain.
- Consider planned/pre-configured sustainment packages of essential items.
- Plan for increased vehicular maintenance especially over rough terrain.
- Maximize maintenance support teams well forward.
- Request distribution at forward locations.
- Increase use of meals-ready-to-eat (MRE).
- Use captured enemy supplies and equipment, and particularly support vehicles and POL. Before use, test for contamination.
- Suspend most field service functions except airdrop and mortuary affairs.
- Prepare thoroughly for casualty evacuation and mortuary affairs requirements.
- Select potential/projected supply routes, logistic release points, and support areas based on map reconnaissance.
- Plan and coordinate enemy prisoner of war operations.
- Plan replacement operations based on known/projected losses.
- Consider the increasing distances and longer travel times for supply operations.
- Ensure that sustainment preparations for the attack do not compromise tactical plans.

## SUPPORTING DEFENSIVE OPERATIONS

9-57. The BCT commander positions sustainment assets to support the forces in the defense and survive. Sustainment requirements in the defense depend on the type of defense. For example, increased quantities of ammunition and decreased quantities of fuel characterize most area defensive operations. However, in a mobile defense, fuel usage may be a critical part of support. Barrier and fortification materiel to support the defense often has to move forward, placing increased demands on the transportation system (FM 4-0).

9-58. The following sustainment techniques and considerations apply to defensive planning:

- Preposition ammunition, POL, and barrier materiel in centrally located position well forward.
- Make plans to destroy those stocks if necessary.
- Resupply during limited visibility to reduce the chance of enemy interference.
- Plan to reorganize to reconstitute lost sustainment capability.
- Use maintenance support teams in the unit maintenance collection point to reduce the need to recover equipment to the brigade support area.
- Consider and plan for the additional transportation requirements for movement of CL IV barrier materiel, mines, and pre-positioned ammunition, plus the sustainment requirements of additional engineer units assigned for preparation of the defense.
- Plan for pre-positioning and controlling ammunition on occupied and prepared defensive positions.

## SUPPORTING STABILITY OPERATIONS IN A HOSTILE ENVIRONMENT

9-59. Sustainment in stability operations involves supporting U.S. and multinational forces in a wide range of missions. Stability operations range from long-term sustainment-focused operations in humanitarian and civic assistance missions to major short-notice peace enforcement missions. Some stability operations may involve combat. Tailoring sustainment to the requirements of a stability operation is key to success of the overall mission (FM 4-0). The sustainment techniques and considerations that are applicable to offensive and defensive operations also apply to stability operations conducted in hostile environment. When these operations are in urban areas, the following considerations may also apply:

- Preconfigure resupply loads and push them forward at every opportunity.
- Provide supplies to using units in required quantities as close as possible to the location where those supplies are needed.
- Protect supplies and sustainment elements from the effects of enemy fire.
- Disperse and decentralize sustainment elements with proper emphasis on communications, command and control, security, and proximity of MSR.
- Plan for extensive use of carrying parties.
- Plan for and use host country support and civil resources when practical.
- Position support units as far forward as the tactical situation permits.
- Plan for special equipment such as rope, grappling hooks, ladders, and hand tools.

This page intentionally left blank.

# Glossary

| | |
|---|---|
| AA | avenue of approach |
| AASLT | air assault |
| ABCS | Army Battle Command System |
| AC2 | airspace command and control |
| ACA | airspace control authority |
| ACE | aviation combat element (USMC) |
| ACO | airspace control order |
| ADAM | air defense and airspace management |
| AFATDS | Advanced Field Artillery Tactical Data System |
| AFSB | Army field support brigade |
| AGM | attack guidance matrix |
| AHS | Army health system |
| ALO | air liaison officer |
| AMD | air and missile defense |
| AMDWS | air and missile defense workstation |
| AO | area of operations |
| ARFORGEN | Army forces generation |
| ASAS | All Source Analysis System |
| ASCOPE | areas, structures, capabilities organizations people, events |
| ASOC | air support operations center |
| ATACMS | Army Tactical Missile System |
| ATGM | antitank guided missile |
| ATO | air tasking order |
| BAE | brigade aviation element |
| BAS | battalion aid station |
| BCCS | Battle Command Common Services |
| BCS3 | battle command sustainment support system |
| BCT | brigade combat team |
| BDA | battle damage assessment |
| BFSB | battlefield surveillance brigade |
| BFT | blue force tracker |
| BHL | battle handover line |
| BHO | battle handover |
| BOLT | brigade operational law team |
| BP | battle position |
| BSA | brigade support area |
| BSB | brigade support battalion |
| BSC | brigade combat company |

| | |
|---|---|
| BSTB | brigade special troops battalion |
| BTB | brigade troops battalion |
| C2 | command and control |
| CA | civil affairs |
| CAB | combined arms battalion |
| CAS | close air support |
| CASEVAC | casualty evacuation |
| CBRN | chemical biological, radiological, and nuclear |
| CBT | combat |
| CCA | close combat attack |
| CCIR | commander's critical information requirement |
| CFV | cavalry fighting vehicle |
| CGS | common ground station |
| CHEMO | chemical officer |
| CIM | civil information management |
| CL | class |
| CMO | civil-military operations |
| CMOC | civil-military operations center |
| CNR | combat net radio |
| COA | course of action |
| COIN | counterinsurgency |
| COLT | combat observation and lasing team |
| COMCAM | combat camera |
| COMSEC | communication security |
| COP | common operational picture |
| COR | contracting office representative |
| CP | command post |
| CPOF | command post of the future |
| CR | civil reconnaissance |
| CRC | control and reporting center |
| CSB | contracting support brigade |
| CSSB | combat sustainment support battalion |
| CTCP | combat trains command post |
| D3A | decide detect, deliver, and assess |
| DCGS-A | Distributed Command Ground System-Army |
| DCO | deputy commanding officer |
| DHIMS | Defense Health Information Management System |
| DLIC | detachment left in contact |
| DOD | Department of Defense |
| DOS | Department of State |
| DP | decision point |

| | |
|---|---|
| **DS** | direct support |
| **DST** | decision support template |
| **DTSS** | Digital Topographic Support System |
| **DZ** | drop zone |
| **EA** | electronic attack |
| **EAB** | echelons above brigade |
| **EGR** | engineer |
| **ENCOORD** | engineer coordinator |
| **EOD** | explosive ordnance disposal |
| **EPB** | electronic preparation of the battlefield |
| **EPLRS** | enhanced position location and reporting system |
| **EPW** | enemy prisoner of war |
| **EQP** | equipment |
| **EW** | electronic warfare |
| **F3EAD** | find, fix, finish, exploit, analyze, and disseminate |
| **FA** | field artillery |
| **FARP** | forward air reserve point |
| **FBCB2** | Force XXI battle command brigade and below |
| **FDC** | fire direction center |
| **FEBA** | forward edge of the battle area |
| **FID** | foreign internal defense |
| **FLD** | field |
| **FLOT** | forward line of own troops |
| **FM** | field manual, frequency modulation |
| **FMI** | field manual interim |
| **FMT** | field maintenance team |
| **FOB** | forward operating base |
| **FRAGO** | fragmentary order |
| **FS** | fire support |
| **FSC** | forward support company |
| **FSCL** | fire support coordination line |
| **FSCM** | fire support coordinating measures |
| **FSCOORD** | fire support coordinator |
| **FSF** | foreign security force |
| **FSNCO** | fire support noncommissioned officer |
| **FSO** | fire support officer |
| **FS3** | fire support sensor system |
| **FTCP** | field trains command post |

| | |
|---|---|
| GCCS-A | Global Command and Control System - Army |
| GIG | global information grid |
| GS | general support |
| HBCT | Heavy Brigade Combat Team |
| HCT | human intelligence collection team |
| HF | high frequency |
| HHB | headquarters and headquarters battery |
| HHC | headquarters and headquarters company |
| HHT | headquarters and headquarters troop |
| HMMWV | high-mobility multipurpose wheeled vehicle |
| HNS | host nation support |
| HPT | high-payoff target |
| HPTL | high-payoff target list |
| HQ | headquarters |
| HR | human resources |
| HSS | health service support |
| HUMINT | human intelligence |
| HVT | high-value target |
| HVTL | high-value target list |
| IBCT | Infantry Brigade Combat Team |
| IE | information engagement |
| IED | improvised explosive device |
| IEW | intelligence and electronic warfare |
| IM | information management |
| IMETS | Integrated Meteorological System |
| INFOSYS | information system |
| INTEG | integration |
| IO | information operations |
| IP | internet protocol |
| IPB | intelligence preparation of the battlefield |
| IR | information requirement |
| ISDN | integrated service digital network |
| ISR | intelligence surveillance, and reconnaissance |
| ISYSCON | integrated system control |
| JAAT | joint air attack team |
| JARN | joint air request net |
| JFC | joint force commander |
| JIIM | joint interagency, intergovernmental, multinational |
| JNN | joint network node |
| JOA | joint operations area |
| JP | joint publication |

| | |
|---|---|
| **JSEAD** | joint suppression of enemy air defense |
| **JSTARS** | Joint Surveillance Target Attack Radar System |
| **JTF** | joint task force |
| **JTRS** | joint tactical radio system |
| **KIA** | killed in action |
| **KM** | knowledge management |
| **LAN** | local area network |
| **LCC** | land component commander |
| **LCE** | logistics combat element |
| **LCMR** | lightweight countermortar radars |
| **LD** | line of departure |
| **LNO** | liaison officer |
| **LOA** | limit of advance |
| **LOC** | line of communication |
| **LOGPAC** | logistics package |
| **LOGSITREP** | logistics situation report |
| **LRP** | logistics release point |
| **LZ** | landing zone |
| **MA** | mortuary affairs |
| **MAGTF** | Marine air-ground task force |
| **MANPADS** | man-portable air defense |
| **MASINT** | measurement and signatures intelligence |
| **MBA** | main battle area |
| **MC4** | medical communications for combat casualty care |
| **MCDP** | Marine Corps Doctrine Publication |
| **MCO** | major combat operations |
| **MCoE** | Maneuver Center of Excellence |
| **MCOO** | modified combined obstacle overlay |
| **MCP** | maintenance collection point |
| **MCS** | maneuver control system |
| **MDMP** | military decision-making process |
| **MEB** | maneuver enhancement brigade |
| **MEDCOM** | medical command |
| **MEDEVAC** | medical evacuation |
| **MEF** | Marine expeditionary force |
| **METT-TC** | (Army only) mission, enemy, terrain and weather, troops and support available, time available, and civil considerations |
| **MEU** | Marine expeditionary unit |
| **MGS** | mobile gun system |
| **MI** | military intelligence |
| **MICO** | military intelligence company |
| **MOB** | mobility |

| | |
|---|---|
| **MOPP** | mission-oriented protective posture |
| **MP** | military police |
| **MRE** | meals-ready-to-eat |
| **MSE** | mobile subscriber equipment |
| **MSR** | main supply route |
| **NAI** | named area of interest |
| **NATO** | North Atlantic Treaty Organization |
| **NBCRV** | nuclear, biological, and chemical reconnaissance vehicle |
| **NCO** | noncommissioned officer |
| **NCOIC** | noncommissioned officer in charge |
| **NDP** | naval doctrine publication |
| **NGO** | nongovernmental organization |
| **NSFS** | naval surface fire support |
| **NTDR** | near term digital radio |
| **OBJ** | objective |
| **OE** | operational environment |
| **OIC** | officer in charge |
| **OP** | observation post |
| **OPCON** | operational control |
| **OPLAN** | operation plan |
| **OPORD** | operation order |
| **OPSEC** | operations security |
| **PA** | public affairs |
| **PAA** | position area for artillery |
| **PAO** | public affairs officer |
| **PBX** | priviate branch exchange |
| **PIR** | priority intelligence requirement |
| **PL** | phase line |
| **PLD** | probable line of departure |
| **PM** | provost marshal |
| **PMESII-PT** | politics, military, economic, social, information, infrastructure, plus physical environment and time |
| **POL** | petroleum, oils, and lubricants |
| **PP** | passage point |
| **PS** | personal staff |
| **PZ** | pickup zone |
| **RF** | radio frequency |
| **ROE** | rules of engagement |
| **RS** | reconnaissance squadron |
| **RSOA** | Ranger, Army Special Operations Aviation |
| **SA** | situational awareness |
| **SALUTE** | size, activity, location, unit, time and equipment |

| | |
|---|---|
| **SATCOM** | satellite communications |
| **SBCT** | Stryker Brigade Combat Team |
| **SCATMINE** | scatterable mine |
| **SCI** | sensitive compartmented information |
| **SEAD** | suppression of enemy air defenses |
| **SEAL** | sea-air-land team |
| **SF** | special forces |
| **SFA** | security force assistance |
| **SGM** | sergeant major |
| **SIGINT** | signals intelligence |
| **SINCGARS** | single channel ground and airborne radio system |
| **SIPR** | Secret Internet Protocol Router Network |
| **SIT** | situation, situational |
| **SITEMP** | situation template |
| **SMART-T** | secure mobile antijam reliable tactical-terminal |
| **SO** | special operations |
| **SOF** | special operations forces |
| **SOP** | standard operating procedure |
| **SPOTREP** | spot report |
| **SPT** | support |
| **SU** | situational understanding |
| **TA** | target acquisition |
| **TAC CP** | tactical command post |
| **TACON** | tactical control |
| **TACP** | tactical air control party |
| **TACSAT** | single-channel tactical satellite |
| **TAI** | targeted area of interest |
| **TAIS** | Tactical Airspace Integration System |
| **TBC** | Tactical Battle Command |
| **TCF** | tactical combat force |
| **TGT** | target |
| **TOC** | tactical operations center |
| **TRP** | target reference point |
| **TSC** | Theater Support Command |
| **TSM** | target synchronization matrix |
| **TSS** | target selection standard |
| **TTP** | tactics techniques and procedures |
| **TUAS** | tactical unmanned aircraft system |
| **UAS** | unmanned aircraft system |
| **UGS** | unattended ground sensors |
| **UMCP** | unit maintenance collection point |

| | |
|---|---|
| **UMT** | unit ministry team |
| **USACE** | U.S. Army Corps of Engineers |
| **USAF** | United States Air Force |
| **UXO** | unexploded ordnance |
| **WARNO** | warning order |
| **WIN-T** | Warfighter Information Network-Tactical |
| **XO** | executive officer |

# References

## SOURCES USED
These are the sources quoted or paraphrased in this publication.

### JOINT PUBLICATIONS
Joint Publications are available online: http://www.dtic.mil/doctrine/index.html.

JP 1-02, Department of Defense Dictionary of Military and Associated Terms, 12 April 2001.
JP 3-0, Joint Operations, 17 September 2006.
JP 3-05, Doctrine for Joint Special Operations, 17 December 2003.
JP 3-08, Interagency, Intergovernmental Organization and Nongovernmental Organization Coordination During Joint Operations, 17 March 2006.
JP 3-09.3, Close Air Support, 8 July 2009.
JP 3-10, Joint Security Operations in Theater, 3 February 2010.
JP 3-18, *Joint Forcible Entry Operations*, 16 June 2008.
JP 4-10, Operational Contract Support, 17 October 2008.

### ARMY PUBLICATIONS
Army Publishing Directorate website: (www.apd.army.mil).

FM 1-0, *Human Resources Support*, 6 April 2010.
FM 1-02, *Operational Terms and Graphics*, 21 September 2004.
FM 2-0, *Intelligence*, 23 March 2010.
FM 2-01.3, *Intelligence Preparation of the Battlefield/Battlespace*, 15 October 2009.
FM 2-19.4, *Brigade Combat Team Intelligence Operations*, 25 November 2008.
FM 3-0, *Operations*, 27 February 2008.
FM 3-01.4, *Multiservice Tactics, Techniques, and Procedures for Joint Suppression of Enemy Air Defenses and Antiradiation Missiles*, 28 May 2004.
FM 3-01.11, *Air Defense Artillery Reference Handbook*, 23 October 2007.
FM 3-04.113, *Utility and Cargo Helicopter Operations*, 7 December 2007.
FM 3-04.126, *Attack Reconnaissance Helicopter Operations*, 16 February 2007.
FM 3-05, *Army Special Operations Forces*, 20 September 2006.
FM 3-05.40, *Civil Affairs Operations*, 29 September 2006.
FM 3-05.401, *Civil Affairs Tactics, Techniques, and Procedures*, 5 July 2007.
FM 3-06.20, *Cordon and Search Multi-service Tactics, Techniques, and Procedures for Cordon and Search Operations*, 25 April 2006.
FM 3-07, *Stability Operations*, 6 October 2008.
FM 3-07.1, *Security Force Assistance*, 1 May 2009.
FM 3-09.21, *Tactics, Techniques, and Procedures for the Field Artillery Battalion*, 22 March 2001.
FM 3-09.70, *Tactics, Techniques, and Procedures for M109A6 Howitzer (Paladin) Operations*, 1 August 2000.
FM 3-13, *Information Operations: Doctrine, Tactics, Techniques, and Procedures*, 28 November 2003.
FM 3-20.96, *Reconnaissance and Cavalry Squadron*, 12 March 2010.
FM 3-21.20, *The Infantry Battalion*, 13 December 2006.
FM 3-21.21, *The Stryker Brigade Combat Team Infantry Battalion*, 8 April 2003.
FM 3-24, *Counterinsurgency*, 15 December 2006.
FM 3-24.2, *Tactics in Counterinsurgency*, 21 April 2009.
FM 3-28.1, *Multiservice Tactics, Techniques and Procedures for Civil Support (CS) Operations*, 3 December 2007.
FM 3-34, *Engineer Operations*, 2 April 2009.
FM 3-34.22, *Engineer Operations – Brigade Combat Team and Below*, 11 February 2009.
FM 3-36, *Electronic Warfare in Operations*, 25 February 2009.
FM 3-39, *Military Police Operations*, 16 February 2010.
FM 3-39.40, *Internment and Resettlement Operations*, 12 February 2010.

FM 3-52, *Army Airspace Command and Control in a Combat Zone*, 1 August 2002.
FM 3-90, *Tactics*, 4 July 2001.
FM 3-90.5, *The Combined Arms Battalion*, 7 April 2008.
FM 3-90.31, *Maneuver Enhancement Brigade Operations*, 26 February 2009.
FM 3-90.61, *The Brigade Special Troops Battalion*, 22 December 2006.
FM 4-0, *Sustainment*, 30 April 2009.
FM 4-02.2, *Medical Evacuation*, 8 May 2007.
FM 4-92, *Contracting Support Brigade*, 12 February 2010.
FM 5-0, *The Operations Process*, 26 March 2010.
FM 5-19, *Composite Risk Management*, 21 August 2006.
FM 6-0, *Mission Command: Command and Control of Army Forces*, 11 August 2003.
FM 6-01.1, *Knowledge Management Section*, 29 August 2008.
FM 6-02.43, *Signal Soldier's Guide*, 17 March 2009.
FM 6-02.53, *Tactical Radio Operations*, 5 August 2009.
FM 6-20, *Fire Support in the Air-Land Battle*, 17 May 1988.
FM 6-20-10, *Tactics, Techniques and Procedures for the Targeting Process*, 8 May 1996.
FM 6-20-40, *Tactics, Techniques, and Procedures for Fire Support for Brigade Operations (Heavy)*, 5 January 1990.
FM 6-20-50, *Tactics, Techniques, and Procedures for Fire Support for Brigade Operations (Light)*, 5 January 1990.
FM 6-22, *Army Leadership Competent, Confident, and Agile*, 12 October 1996.
FM 7-0, *Training for Full Spectrum Operations*, 12 December 2008.
FM 7-15, *The Army Universal Task List*, 27 February 2009.
FM 90-4, *Air Assault Operations*, 16 March 1987.
FMI 2-01, *Intelligence, Surveillance, and Reconnaissance (ISR) Synchronization*, 11 November 2008.
FMI 4-93.2, *The Sustainment Brigade*, 4 February 2009.
FMI 4-93.41, *Army Field Support Brigade Tactics, Techniques, and Procedures*, 22 February 2007.
FMI 6-02.45, *Signal Support to Theater Operations*, 5 July 2007.
TC 1-400, *Brigade Aviation Element Handbook*, 27 April 2006.

# READINGS RECOMMENDED

These sources contain relevant supplemental information.

## ARMY PUBLICATIONS

AR 190-8, *Enemy Prisoners of War, Retained Personnel, Civilian Internees and Other Detainees*, 1 October 1997.
AR 27-1, *Legal Services, Judge Advocate Legal Services*, 30 September 1996.
AR 40-3, *Medical, Dental, and Veterinary Care*, 22 February 2008.
AR 40-400, *Patient Administration*, 27 January 2010.
FM 1, *The Army*, 14 June 2005.
FM 3-06, *Urban Operations*, 26 October 2006.
FM 3-06.11, *Combined Arms Operations in Urban Terrain*, 28 February 2002.
FM 3-07.31, *Peace Operations Multi-Service Tactics, Techniques, and Procedures for Conducting Peace Operation*, 26 October 2003.
FM 3-09.32, *JFIRE Multiservice Tactics, Techniques, and Procedures for the Joint Application of Firepower*, 20 December 2007.
FM 3-09.34, *Multiservice Tactics, Techniques, and Procedures for Kill Box Employment*, 4 August 2009.
FM 3-11, *Multiservice Tactics, Techniques, and Procedures for Nuclear, Biological, and Chemical Defense Operations*, 10 March 2003.
FM 3-11.19, *Multiservice Tactics, Techniques, and Procedures for Nuclear, Biological, and Chemical Reconnaissance*, 30 July 2004.
FM 3-19.4, *Military Police Leaders' Handbook*, 4 March 2002.
FM 3-34.2, *Combined-Arms Breaching Operations*, 31 August 2000.
FM 3-34.210, *Explosives Hazards Operations*, 27 March 2007.

FM 3-34.400, *General Engineering,* 9 December 2008.

FM 3-37, *Protection,* 30 September 2009.

FM 3-50.1, *Army Personnel Recovery,* 10 August 2005.

FM 3-52.1, *Multiservice Tactics, Techniques, and Procedures for Airspace Control,* 22 May 2009.

FM 3-52.2, *TAGS Multiservice Tactics, Techniques and Procedures for the Theater Air Ground System,* 10 April 2007.

FM 3-90.1, *Tank and Mechanized Infantry Company Team,* 9 December 2002.

FM 3-90.12, *Combined Arms Gap-Crossing Operations,* 1 July 2008.

FM 3-90.119, *Combined Arms Improvised Explosive Defeat Operations,* 21 September 2007.

FM 4-01.30, *Movement Control,* 1 September 2003.

FM 4-02, *Force Health Protection in a Global Environment,* 13 February 2003.

FM 4-02.1, *Army Medical Logistics,* 8 December 2009.

FM 4-02.4, *Medical Platoon Leader's Handbook Tactics, Techniques, and Procedures,* 24 August 2001.

FM 5-34, *Engineer Field Data,* 19 July 2005.

FM 6-22.5, *Combat and Operational Stress Control Manual for Leaders and Soldiers,* 18 March 2009.

FM 8-42, *Combat Health Support in Stability Operations and Support Operations,* 27 October 1997.

FM 90-26, *Airborne Operations,* 18 December 1990.

FMI 2-01.301, *Specific Tactics, Techniques, and Procedures and Applications for Intelligence Preparation of the Battlefield,* 31 March 2009.

## JOINT PUBLICATIONS

JP 3-13, *Information Operations,* 13 February 2006.

JP 3-28, *Civil Support,* 14 September 2007.

JP 3-34, *Joint Engineer Operations,* 12 February 2007.

JP 3-57, *Civil-Military Operations,* 8 July 2008.

JP 3-63, *Detainee Operations,* 30 May 2008.

## OTHER

DODD 5210.56, Use of Deadly Force and the Carrying of Firearms by DOD Personnel Engaged in Law Enforcement and Security Duties, 1 November 2001.

DODD 5525.5, DOD Cooperation with Civilian Law Enforcement Officials, 15 January 1986.

DODD 2000.15, Support to Special Events, 21 November 1994.

Marine Corps Doctrine Publication [MCDP] 3, Expeditionary Operations, 16 April 1998.

Naval Doctrine Publication [NDP] 1, Naval Warfare, 28 March 1994.

GTA 19-07-001, *Enemy Prisoner of War (EPW) Basic Commands,* 1 February 1989.


## DEPARTMENT OF THE ARMY FORMS

DA Forms are available on the Army Publishing Directorate website (www.apd.army.mil).

DA Form 2028, *Recommended Changes to Publications and Blank Forms.*

DA Form 3953, *Purchase Request and Commitment.*

This page intentionally left blank.

# Index